# Patient Handling in the Healthcare Sector

A Guide for Risk Management with
MAPO Methodology (Movement and
Assistance of Hospital Patients)

# Patient Handling in the Healthcare Sector

A Guide for Risk Management with
MAPO Methodology (Movement and
Assistance of Hospital Patients)

Olga Menoni • Natale Battevi • Silvia Cairoli

CRC Press
Taylor & Francis Group
Boca Raton   London   New York

CRC Press is an imprint of the
Taylor & Francis Group, an **informa** business

international
ergonomics
school

Parts of this book were previously published in Italian: *The MAPO Method for the Analysis and Prevention of Risk from Handling Patients;* published by FrancoAngeli s.r.l., Milano, Italy.

CRC Press
Taylor & Francis Group
6000 Broken Sound Parkway NW, Suite 300
Boca Raton, FL 33487-2742

© 2015 by Taylor & Francis Group, LLC
CRC Press is an imprint of Taylor & Francis Group, an Informa business

No claim to original U.S. Government works

Printed on acid-free paper
Version Date: 20140617

International Standard Book Number-13: 978-1-4822-1718-6 (Paperback)

**Library of Congress Cataloging-in-Publication Data**

Menoni, Olga, author.
    Patient handling in the healthcare sector : a guide for risk management with MAPO methodology (movement and assistance of hospital patients) / Olga Menoni, Natale Battevi, and Silvia Cairoli.
    p. ; cm.
    Guide for risk management with movement and assistance of hospital patients methodology
    Includes bibliographical references and index.
    ISBN 978-1-4822-1718-6 (alk. paper)
    I. Battevi, Natale, author. II. Cairoli, Silvia, author. III. Title. IV. Title: Guide for risk management with movement and assistance of hospital patients methodology.
    [DNLM: 1. Moving and Lifting Patients--adverse effects. 2. Occupational Injuries--etiology. 3. Low Back Pain--prevention & control. 4. Occupational Injuries--prevention & control. WY 100.2]
    RC965.M39
    363.15--dc23                                                                                      2014012590

**Visit the Taylor & Francis Web site at**
**http://www.taylorandfrancis.com**

**and the CRC Press Web site at**
**http://www.crcpress.com**

# Contents

# Preface

It is widely recognized that the operators responsible for the care of patients, regardless of the title of their profession (nurses, nurse's aides, etc.), constitute a working population exposed to biomechanical overload risk, particularly of the spine and shoulder.

This risk factor rises and will continue to rise going forward with the progressive aging of the population because the number of patients who need assistance is steadily increasing.

As stated by the European Commission, the number of workers who are unfit for handling patients is destined to rise, and this will have a negative impact on the management of personnel.

This handbook is devoted to all those operating in the healthcare sector who carry out manual patient lifting and handling tasks, especially healthcare managers and workers, occupational safety and health caregivers, manufacturers of assistive devices and equipment, education and training supervisors, and designers of healthcare facilities (according to the criteria laid down by international standards).

It goes without saying that parametric approaches toward risk assessment, such as the one presented in this book, are not always welcomed with open arms, as they inevitably overlook certain aspects. We hope that the characteristics of our method will enable it to be regarded as holistic, and that readers will appreciate the focus we place on the organization of work.

Our aim is to propose an approach that will allow certain objectives to be reached:

- Rapid analysis of the problem
- Rapid identification of solutions
- Effective monitoring of the effects of preventive actions

As for other observational risk assessment methods, one of the features of our approach is that it employs tools that allow those responsible for allocating financial resources to estimate what investments need to be made to achieve specific results. This means taking the decision-making process out of the hands of ergonomics experts and putting it into those of healthcare facility administrators.

# Acknowledgments

It is not easy to thank all those whose countless contributions enabled this book to be written, but we will do our best not to overlook anyone.

Driving us relentlessly forward as we were putting together this book, in English, were our colleagues Enrico Occhipinti and Daniela Colombini, with whom we have long shared a passion for researching work-related biomechanical overload risk. Our relationship goes far beyond a mere coming together of ideas.

Special thanks go to Marco Tasso: Our young colleague oversaw the review process and contributed directly to clarifying many, sometimes quite complex, concepts. We are indebted to all those who helped us broaden our outlook, which began with a primarily Italian focus and now encompasses other organizations and cultures: Enrique Alvarez Casado, Sonia Tello Sandoval, Diana Robla, Begona Baiget, Aquilles Hernadez, Rudy Facci, and Edoardo Santino.

Over the past 15 years we have had the great fortune to meet and interact with many people who have devoted their talents and skills to exploring this subject within the framework of a working group created on a voluntary basis in 2005 called EPPHE (European Panel on Patient Handling Ergonomics). The group also contributed to the drafting of ISO TR 12296. We would like to single out Sue Hignett, Mike Fray, Elly Waijer, Hanneke Knibbe, Leena Tamminen, and Matthias Jaeger.

The MAPO (*movimentazione e assistenza pazienti ospedalizzati*—movement and assistance for hospitalized patients) methodology described in this book is the result of close cooperation with numerous Italian colleagues, whom we thank wholeheartedly: Paola Torri, Giorgio Zecchi, Alberto Baratti, and Rosa Manno, Mariaadelia Rossi, Nora Vitelli, Flavio Verona, Daniela Panciera, and MariaGrazia Ricci.

# The Authors

**Olga Menoni** is an ergonomist and physiotherapist at the Center for Occupational Medicine (CEMOC) at the Foundation Ca' Granda Policlinico in Milan, Italy. She received her degree in occupational physiotherapy in 1979 and has been an ergonomist since 2000. She served as senior researcher for the EPM (Ergonomics of Posture and Movement) Research Unit as a result of an agreement among Milan University, the Policlinico Foundation, and the Don Gnocchi Foundation, and is a professor of the science of prevention at the University of Milan. She has been a professor for more than 100 training courses for occupational physicians and technicians in the management of lifting patients. She has authored more than 50 papers and handbooks, in Italian and English, on occupational health and ergonomics issues, with special focus on prevention of work-related musculoskeletal disorders in the hospital caused by manually handling patients. She is a member of the Technical Committee on the Prevention of Musculoskeletal Disorders of the International Ergonomics Association (IEA).

**Natale Battevi** holds a degree in medicine (1978) and attended postgraduate school in occupational health in 1982 and health statistics in 1989. From 1997 to 2012, she worked with the ergonomics section of the CEMOC, Clinica del Lavoro, at the Foundation Ca' Granda Policlinico in Milan. Since 1997 she has been an integral part of the EPM Research Unit as a result of an agreement among Milan University, the Policlinico Foundation, and the Don Gnocchi Foundation. Active in CEN and ISO international standards of ergonomics, she has served as an ergonomics consultant for many national companies and for national health services. She has worked as a professor in the science of prevention at the University of Milan and has authored numerous papers and handbooks, in Italian and English, on occupational health and ergonomics issues, with a special focus on the prevention of work-related musculoskeletal disorders.

**Silvia Cairoli** received a degree in medicine in 1988, followed by study of occupational health at postgraduate school in 1992. From 1999 to 2012 she worked with the CEMOC ergonomics section, Clinica del Lavoro, at the Foundation Ca' Granda Policlinico in Milan. Since 1999 she has been an integral part of the EPM Research Unit as a result of an agreement among Milan University, the Policlinico Foundation, and the Don Gnocchi Foundation. She has authored numerous papers and handbooks, in Italian, on occupational health and ergonomics issues, with a special focus on the prevention of work-related musculoskeletal disorders.

# 1 Introduction

## A Brief History of the MAPO Index Equation and Relationship with a Specific Preventive Plan

In the Italian and international scientific literature, healthcare workers involved in caring for dependent patients are among those most prone to acute and chronic musculoskeletal disorders, especially the dorsolumbar spine. This is consistent with the data from countless investigations showing that the manual handling of noncooperative patients is associated with substantial overload of the lumbar spine, often far exceeding limits deemed to be "physiological."

According to international data (WHO 2011), there are around 19,300,000 nursing staff members working in healthcare facilities around the globe; about 85% are women.

Most healthcare workers manually handle patients on a daily basis, an activity that Italian and European regulations consider "potentially" hazardous in terms of workplace risk prevention, thus requiring hospital management to adopt effective risk assessment, management, and containment measures.

However, it is not only regulations and standards that are driving the need for action to be taken. For instance, it is worth noting that a large percentage of nursing staff (from 6% in hospitals to 13% in geriatric facilities) are affected by musculoskeletal disorders, preventing them from performing activities for manually handling patients. As well, the level of sick days taken due to musculoskeletal problems (acute low-back pain) is extremely high, and increasing numbers of healthcare workers are realizing that their back problems are work related. In complex organizations like hospitals, all these aspects (together with many others) raise inefficiency, generate higher costs, and, ultimately, lower the quality of care.

Therefore, the problem must be "managed" in the broadest sense, bearing in mind the twofold aim of safeguarding worker health and ensuring quality care for patients and the population at large.

The scientific community in general and healthcare workers in particular (at both the Italian and international levels) are well aware of these issues, as evidenced by the abundance of articles, reviews, guidelines, and action plans offering recommendations, ideas, experiences, and efforts to address the problem, albeit in countries where conditions vary quite dramatically.

Only recently has there been a clear convergence between leading experts and nursing staff representatives with respect to preventive measures that must be based on a comprehensive healthcare sector-wide strategy capable of tackling the many aspects influencing risk associated with handling and caring for dependent patients: organizational (i.e., staffing levels, definition of care-giving procedures, work duration, work relationships), technical/structural (i.e., availability and quality of aids, types and organization of wards and care), and educational (i.e., adequate training and understanding of care-giving procedures and techniques).

Over the long term, it has been proven that only broadly based strategies are capable of adequately managing worker risk—reducing injury, sickness, absenteeism, costs, and, ultimately, leading to higher quality patient care. Based on ample evidence in the international literature, partial interventions, such as training alone or the mere provision of mechanical aids, are largely ineffective, often defeating the purpose of well-intentioned financial investments.

One of the drivers of a comprehensive intervention strategy is undoubtedly the risk assessment. While risk assessments nominally serve to estimate the degree of biomechanical overload for the musculoskeletal system and the resulting likelihood of injury due to handling patients manually, they should—especially in this field—be regarded as a detailed analysis of every aspect affecting risk (i.e., organizational, technical, structural, and educational) as well as for the purposes of identifying any criticalities that could be minimized or eliminated. An assessment capable of using various methods, including parameters, to gauge all risk factors represents the best approach for ultimately targeting and reducing risk. This argument is supported by a reliable source: the ISO 12296 Technical Report (TR) published on June 1, 2012: "Ergonomics—Manual Handling of People in the Healthcare Sector" specifically focused on handling patients.

But above and beyond these details, the aforesaid TR bridges a specific gap in the overall international standards governing physical ergonomics, speaking to the need for a strategic approach to the problem based on an adequate risk assessment to guide integrated and interdisciplinary improvements. On the one hand, after an in-depth review, the TR suggests and reports on certain internationally recognized methods for carrying out global, analytical, and parameter-based risk assessments (including the Italian MAPO method); on the other, via the inclusion of extensive annexes, it recommends an integrated approach to risk management in terms of organizational aspects, the provision of sufficient and suitable mechanical aids, and the logistics of healthcare facilities, operator training, and ongoing monitoring of the effectiveness of all corrective actions. The technical report is explicitly directed toward all those involved in managing the specific issue: healthcare managers, healthcare workers, prevention officers, trainers, handling aid manufacturers, and healthcare facility designers.

The authors of this book, alongside their mainly European colleagues, were closely involved in drafting this TR and unquestionably shaped and shared its underlying concepts and content, based on their experience in Italy over the past 15 years. Based also on recent findings, the authors provide a detailed description of the Italian method for assessing patient-handling risk (the acronym is MAPO— *movimentazione e assistenza pazienti ospedalizzati*—movement and assistance for hospitalized patients). They developed this, after more than 15 years of use, for the integrated assessment and management of risk associated with the many and varied activities involved in handling patients in healthcare settings.

The method dates back to around 1995 and includes detailed data concerning approximately 200 in-patient wards located mainly in northern Italy. The focus was primarily on how the work was organized at the time, in hospital environments staffed mainly by nurses rotating among three different shifts, with an average patient stay of 7 days. It was thus possible to identify homogeneous groups of workers in terms of risk exposure in each individual structure (e.g., ward, surgical block, etc.). At the end of the observational phase, the methodology was developed with a view to simplifying the assessment. Two main risk factors were identified: one related to the organization of work and the other to the work environment and equipment or aids used for patient handling.

Over the years, the healthcare sector has undergone dramatic changes: Average hospital stays have become considerably shorter (the average length of hospital stays today is 4 days) with more and more hospitals catering to acute patients and an increasing number of long-term care facilities. This trend is particularly significant in the West, where the populations are aging most rapidly. Another three important developments have also arisen: the increase in non-nursing staff ("nursing aides"), changes to working hours, and the appearance of mechanical aid systems. The method has gradually incorporated the new data, but the underlying rationale has been preserved.

One of the characteristic features of the MAPO method is that it combines all the various risk factors into a single formula—not only to assign a value for the level of risk in a given area or ward, but also to supply valuable recommendations for putting together a preventive action plan taking into account the specific constraints of the structure under examination. While a method based on parameters, such as the MAPO method, does introduce a certain amount of simplification, it arguably offers the benefit of being a practical tool enabling the effectiveness of the proposed improvements to be monitored over the course of time.

Lastly, it should be stressed that the MAPO approach to assessing and preventing risk is based on the typically global, interdisciplinary, and participatory principles of ergonomics. This may also explain why the MAPO method has been used by so many Italian hospitals and healthcare facilities, especially in the northern and central regions of the country, and why it is becoming very popular in the Iberian Peninsula and Spanish- and Portuguese-speaking Latin America.

This wealth of experience, along with many publications over the years, has enabled the method to be enhanced, enriched, and adapted to the specific needs of many healthcare environments and professionals.

Readers are offered a methodological approach that nonetheless focuses much attention on real-life applications, including plentiful operational recommendations and practical tools, both paper based and digital. Like the ISO technical report mentioned earlier, this book was written for many healthcare professionals: hospital managers and directors, nurse directors, healthcare workers in general, risk prevention officers, occupational health physicians (protection and prevention services), and trade union officers in the healthcare environment, as well as healthcare trainers and training bodies. However, the book might also be of interest to healthcare procurement departments, mechanical aid producers, and facility designers and planners.

The first few chapters of the book provide a framework for understanding risk for handling patient manually, after which the reader's attention is drawn to the tools for making detailed risk assessments in different kinds of facilities and areas (e.g., acute and chronic care wards, residential healthcare facilities, surgical blocks, day hospitals and clinics, and emergency departments). Theoretical and methodological sections are supported by real examples and actual situations. Special emphasis is placed on the selection and provision of mechanical aids (patient lifts, beds, gurneys, wheelchairs, and "minor" aids) and on the principal requirements and features of effective healthcare worker training and education. These two last aspects are a *sine qua non* for adequate risk prevention and high-quality patient care; however, their effectiveness and usefulness must be constantly monitored.

This manual provides practical and accessible information for all those involved in the prevention of patient-handling risk in the healthcare sector. We hope you will find it both enjoyable and useful.

# 2 A Review of Work-Related MSDs in Caregivers

## 2.1 INTRODUCTION

Generally speaking, musculoskeletal disorders (MSDs), especially chronic degenerative spinal problems, are very common among the general population: Backache, or low-back pain, is the most frequent complaint. According to the literature, between 60% and 90% of the population suffers from back pain at some time during their lives, and between 15% and 42% have more than one episode during their lifetime. Intense musculoskeletal pain accounts for between 15% and 20% of doctor visits. These widely ranging percentages are due to the different criteria used for choosing sample study populations and, above all, to the different definitions given to the symptom known as "backache." Fortunately, most patients return to health after just one episode of back pain (between 60% and 70% within 6 weeks and between 70% and 90% within 12 weeks); however, sick leave due to backache also determines major social and economic costs because the recurrence rate is extremely high, ranging from 20% to 40% in 1 year and up to 85% during an entire lifetime (OSHA EU, FACTS 2000; OSHA EU 2004, 2007).

Despite significant strides in understanding this pathology, low-back pain is a critical issue that continues to defy solution, hence the relentless rise in disability due to lumbago, which generates very high personal and social costs associated with diagnostic and therapeutic interventions and with the limitations to people both in and out of the workforce.

Recent data published by the Global Burden Disease Study (2013) state that between 1990 and 2010, the main cause of years lived with disability (YLDs) is low-back pain. From the same sources, the WHO announced in 2009 that "37% of back pain can be attributed to occupational risk factors" and is the "major cause of work absences, resulting in economic loss." Consequently, there is an increasing need to identify highly sensitive and specific risk measurement tools in order to design preventive strategies of proven efficacy and efficiency, avoiding a useless waste of resources (Waters 2007; OSHA EU 2007).

In addition to low-back pain, the now somewhat dated literature (Silverstein et al. 2008; Harkness et al. 2003; Hoozemans et al. 2002) reports that among the most relevant musculoskeletal disorders associated with the lifting of weights, the shoulder is undoubtedly the joint most urgently in need of preventive action from the technical, organizational, and healthcare standpoints (Trinkoff et al. 2006; Smedley et al. 2003; Maso et al. 2003; Harkness et al. 2003; Devereux et al. 2001; Vasseljen 2001; Miranda et al. 2008; Hoozemans et al. 2002; Viikari-Juntura 2001; Croft et al.

2001; Silverstein et al. 2008; Van der Windt et al. 2000; Ariens et al. 2000; Palmer et al. 2001).

## 2.2  DEFINITIONS

Most backache sufferers can point to a mechanical cause for their pain; only 2% of causes are visceral (menstrual pain, peptic ulcer or pancreatitis, urinary tract infections, etc.), and 1% is due to nonmechanical causes. Some back problems, although of multifactorial origin, may stem primarily from biomechanical overload or other concomitant causes, such as those listed here (Colombini et al. 2013):

- Degenerative spine diseases
- Lumbar disk disease: loss of disk height with end plate degeneration
- Multiple-level lumbar disk degeneration
- Bulging lumbar disk with impression of the dural sac
- Lumbar disk herniation

Clinically, these diseases and disorders all begin with lumbar pain, but studying them and comparing the results of the various investigations reported in the literature are made all the more difficult due to the different approaches used for choosing sample populations and for defining what is meant by back pain (Colombini et al. 1999).

Conversely, pain radiating to the legs and disk disease have been defined as (PNLG 9 2005):

- **Lumbosciatica:** back pain radiating down the leg and below the knee (due to involvement of L5 or S1, in over 90% of subjects with radiculopathy)
- **Lumbocrural sciatica:** pain radiating to the anterior aspect of the thigh and leg (due to involvement of L2, L3, L4)
- **Herniated disk:** rupture of the annulus fibrosus wall, releasing nucleus material into the spinal canal, generally in the posterior or posterolateral part of the disk
- **Extrusion:** when the hernia pushes through the posterior longitudinal ligament (with possible intervertebral or intraforaminal sequestration, migration, and herniation)
- **Protrusion** (or bulging disk)**:** if the hernia is constrained by the posterior longitudinal ligament

But here, too, there is no consensus in the literature as to the definitions of extrusion or protrusion.

Conversely, with regard to evaluating working populations, at the international level there is extensive consensus on the use of the term **low-back pain (LBP)**, which is defined as follows (OSHA EU, FACTS 2000; Marras 2008): "lumbar pain, which may be related to problems of the spine, vertebral disks, ligaments and muscles, spinal cord and peripheral nerves."

Most studies investigating the prevalence of low-back pain in different working populations, both in Italy and internationally, have focused on **acute lumbar pain**

**TABLE 2.1**
**Low-Back Pain**

| Definition | Symptoms | Duration |
|---|---|---|
| Acute low-back pain | An episode of intense pain in the lumbosacral area, so as to prevent the sufferer from flexing, bending, or rotating the spine (low-back pain or lumbago); onset may be acute or insidious | At least 2 days of bed rest (or 1 day with pain medication) |

as the referring clinical complaint in order to gather data on the dissemination of the problem, as well as on the relationship between injury and level of exposure.

It is for this reason that it appears advisable to continue using the definition as shown in Table 2.1, which effectively links expected lumbar damage to the different workplace exposure levels measured. In the study (and surveillance) of musculoskeletal disorders, reported symptoms are of the utmost importance since they generally have an early onset and, if detected adequately, can suggest a diagnostic suspicion and the need for the most appropriate clinical and/or instrumental exams.

In *Manual Lifting: A Guide to the Study of Simple and Complex Lifting Tasks* (Colombini et al. 2013), an entire chapter is devoted to monitoring the health of workers exposed to biomechanical overload.

## 2.3 A BRIEF OVERVIEW OF THE PATHOGENESIS

The vertebral column, also known as backbone or spine, is a complex structure consisting of a series of bones stacked up like blocks on top of one another to support the body's weight and enable its movement, as well as to protect the nerves enclosed within the spinal cord.

The spine is very strong and can withstand considerable loads; it is divided into an anterior column comprising stacked vertebral bodies, whose function is primarily for weight bearing, and a posterior pillar composed of articular processes and their respective capsular ligaments, which serve for movement. Vertebral disks not only connect the vertebral bodies to one another, but also absorb and distribute mechanical stress during movement. The closer one gets to the lumbosacral region, the greater is the level of applied force. Because of the way it is constructed (i.e., a gelatinous core with an external fibrous ring), the intervertebral disk is able to withstand considerable loads, becoming deformed and then returning to its original size and shape once the mechanical stress has ceased.

However, an excessive load on the spine may cause the tolerance limits of tissues to be exceeded, either acutely or due to cumulative trauma (i.e., low but frequent loading), determining the start of a succession of lesions that may lead to low-back pain and intervertebral disk degeneration (EWCS 2007; Marras 2008; Seidler et al. 2009).

Negative ergonomic factors present in the workplace may add to the normal degeneration process, due to aging, and contribute to low-back problems in an otherwise healthy spine or speed up the changes to an already injured back.

With regard to cumulative trauma, tissue tolerance levels diminish with age, and genetic factors are undoubtedly important in defining muscle recruitment and response to effort; however, while it may be difficult to separate individual from occupational risk factors, the correlation between work and lumbar pain has been well proven. It has been estimated that around two-thirds of all cases of low-back pain are due to work-related causes, especially the manual movement of loads/patients; in certain circumstances, the disk load may be well over tolerance limits (approx. 275 kg for women and 400 kg for men). This roughly corresponds to the *action limit* identified by the National Institute for Occupational Safety and Health (NIOSH) in 1981 (340 kg). Moreover, there are certain movements (i.e., actual lifting actions), in which the load on the intervertebral disk exceeds breaking point, identified as the maximum limit by NIOSH at 650 kg (Marras 2008; Menoni et al.1999).

Lumbar pain is a *perceived* rather than a felt pain and as such may also be influenced by cognitive processes. Although the perception of pain generally stems from physical injury to a biological system, it has now been proven that it occurs within a multidimensional and interactive nervous system; pain can, in fact, continue to be perceived even without a chronic peripheral stimulus, and even after initially damaged tissue has healed. Thus, external and internal factors may interact in ways not yet clearly understood; the recent interpretation offered by Marras (2008) might be the most convincing so far.

### 2.3.1  RISK FACTORS

Going on to occupational risk factors, the NIOSH publication released in 1997 examines numerous epidemiological studies in the literature that indicate a relationship between work and lumbar pain. Following a thorough review based on rigorous epidemiological criteria, NIOSH assessed the potentially significant role (strong evidence) of lifting actions and vibrations transmitted to the whole body, followed (evidence) by strenuous physical work and awkward postures.

The stress endured by the lumbosacral spine during the manual handling of loads can thus be attributed to three main parameters:

- Weight
- Frequency
- Handling method

Other studies (Andersson 1999; Hoogendoorn et al. 1999, 2002; NCHS 2006; Atlas et al. 2004; NRC 2001) have confirmed the role of these risk factors and singled out certain workplaces (manufacturing, transportation, and services) and jobs (retail sales, construction, and healthcare) as being at higher risk.

The aforementioned NIOSH review, as well as countless other studies (Bongers et al. 1993; Toivanen, Helin, and Hanninen 1993; Camerino et al. 1999, 2001, 2004; Marras et al. 2000; Violante et al. 2004; Yip, Ho, and Chan 2001; Hoogendoorn et al. 2002; Harkness et al. 2003; Linton 2000, 2001; Iles, Davidson, and Taylor 2008; Mehlum et al. 2008; Plouvier et al. 2009; Hooftman et al. 2009), also focused attention on the psychological, social, and organizational aspects of the problem.

Poor social support; boring, monotonous, and demotivating work; and excessively high perceived demands may also constitute risk factors for low-back pain. The line separating psychosocial and physical factors in the workplace is blurred, especially when they overlap, although physical risk factors still appear to play a preeminent role (Marras 2008).

Equally interesting are the ideas put forward in 2008 by Kompier and van der Beek, who suggested combining psychosocial risk factors and musculoskeletal disorders, as well as implementing "objective" research methods such as those adopted in 2008 by Elfering and colleagues. They proposed taking increased urinary excretion of catecholamines as an indicator of sympathetic nervous system activation, which could play a significant role in the interaction between work-related stress and musculoskeletal pain.

Even the European Agency for Safety and Health at Work (2007a) set out to come up with a new global approach to the prevention of musculoskeletal pathologies, fostering initiatives for reducing all the physical, individual, and psychosocial factors that may increase or cause them (Table 2.2).

More recently, Marras et al. (2006) looked at biomechanical studies, based on which they stressed that the faster certain actions are performed, the greater is the risk of a progressive and dangerous increase in the intra-abdominal pressure and tangential forces acting on the disk, which may reach or even exceed the tolerance levels determined by the viscoelastic properties of the disk itself. It has been estimated that during a typical 8-hour shift, a nurse may handle a total weight of up to 1.8 tons and that pulling and pushing actions are frequently at high risk.

Similar findings were reported by Jager et al. (2007) in relation to the DOLLY project (Dortmund Lumbar Load Study: a complex research project designed to identify the lumbar load associated with certain specific manual patient-handling activities). DOLLY analyzed 162 patient-handling tasks, estimating the different disk load during handling operations performed using "conventional," "optimized," and "aided" methods. Factoring in the age and gender of the operators as well, the research showed that the lumbar disk load is consistently high and that many activities exceed the recommended limits, especially among older workers. Moreover, it was proven that correctly planning the action and the use of aids may reduce disk loading. Risk levels can also be lowered by training and the use of aids (as demonstrated by Waters, Nelson, and Proctor in 2006) in patient-handling activities carried out in critical care settings. Since the pathologies are multifactorial, individual

## TABLE 2.2
## Work Factors That Increase the Risk of Low-Back Damage

| Physical Factors | Psychosocial Factors | Organizational Factors |
|---|---|---|
| Heavy physical labor | Poor social support | Poor work organization |
| Lifting and handling loads | Monotonous activities | Demotivating jobs |
| Improper positions (lifting weights, twisting upper body, and static positions) | | |
| Vibrations through the whole body (truck driving) | | |

risk factors cannot be underestimated (Ferguson and Marras 1997; Hooftman 2004; Battié 1990; Cady, Thomas, and Karwasky 1985; Miranda 2002; Marras et al. 2000).

The age of onset is generally between 35 and 55, while the number of sick days taken increases with age. There appears to be a positive correlation between age and spinal instability, confirming that muscular control and individual risk both change over the course of time.

Females of all ages are more prone to disk damage, but if work is added to the equation, men seem to be at higher risk, perhaps due to the different ways in which they tend to handle loads. Height (taller than 180 cm for men and 170 cm for women) and **body mass** (BMI > 30) are also important; exercise may instead reduce the incidence of low-back pain and the duration of symptoms. Genetic and familial factors, together with age and workload, would appear to play a major, although not yet well-defined role.

## 2.4   MAGNITUDE OF THE PROBLEM

From the socioeconomic standpoint, low-back pain represents the principal and most extensive cause of spending among all musculoskeletal problems in the industrialized world; this is why it was given the highest priority by the National Institute for Occupational Safety and Health in planning all of its targets in 2000.

In both the United States and Europe, the costs and expenditures associated with work-related musculoskeletal disorders (WMSDs) have a massive impact on budgets, employment strategies, and socioeconomic policies. These disorders are the cause of frequent and prolonged absences from work; estimates speak of an average of 28.6 sick days per 100 workers in the United States, and an even higher level in Scandinavia (36 days off per 100 workers) and in the UK (32.6 days). Similar figures have emerged worldwide from a United Nations-sponsored study conducted by Punnett et al. in 2005.

As early as 2000, the European Agency for Health and Safety at Work chose musculoskeletal disorders as the theme for its "European Week" information campaign. It is no surprise that in 2007 the theme was again adopted for the eighth European Health and Safety at Work Campaign, given how important it has become throughout Europe (European Foundation for the Improvement of Living and Working Conditions 2005).

Musculoskeletal disorders have now well and truly been deemed a priority in Europe, among member states (including the latest new entries), and at the level of European social partners. For certain employment sectors the priority is even higher, such as in healthcare, together with the more traditional building construction, transportation, and service sectors.

Work-related diseases and disorders have been estimated to account for between 2.6% and 3.8% of Europe's GDP. Of that, 50% is caused by musculoskeletal disorders. This explains why the European Agency concluded its policy document by stating that the prevention of such diseases and disorders is undoubtedly "good business."

These general data document the spread of both exposure and damage, confirming the need to continue assessing and preventing risk due to the manual handling of loads.

## 2.5 EPIDEMIOLOGICAL STUDIES ON CLINICAL EFFECTS AMONG HEALTHCARE WORKERS ASSIGNED TO MANUAL PATIENT-HANDLING TASKS

### 2.5.1 THE INTERNATIONAL EXPERIENCE

From the epidemiological standpoint, countless investigations have been conducted to define the risk of developing painful osteoarticular symptoms, especially in the lumbosacral region of the spine, among healthcare workers and caregivers (Magora 1970; Stobbe et al. 1988; Yassi et al. 1995; Knibbe and Knibbe 1996; Smedley et al. 1998, 2003; Lagerstrom, Hansson, and Hagberg 1998; Menoni et al. 1999; Hignett 2001; Engkvist 2000; Smith et al. 2003).

In the literature there are numerous studies looking into the frequency and main factors underlying chronic lumbar degenerative disk disease among healthcare workers.

With regard to finding low-back pain in nursing populations, Stubbs and colleagues, as early as 1983, examined 3,912 nurses working in the main departments of four British hospitals (Psychiatry, Geriatrics, Internal Medicine, Surgery, Ob/Gyn, and others), selected proportionally in such a way as to reproduce the same distribution as in the overall national nursing population. The incidence and period prevalence of low-back pain were, respectively, 77 and 431 per 1,000 at risk. The same research also estimated that of the 430,000 nurses working in the United Kingdom at the time, some 40,000 suffered from episodes of low-back pain every year, causing the loss of a total of 764,000 work days annually.

A review carried out by Hignett in 1996 examined 80 studies of various types (experimental, epidemiological, and prevention related) and evidenced a positive correlation between the work performed by nurses and the presence of low-back pain. Moreover, it has been possible to estimate a point prevalence of 17% and an annual prevalence of 40%–50%, with the incidence increasing as the patient manual handling activities increased.

In 1998, Smedley and colleagues published the results of a longitudinal study involving 906 nurses; a series of questionnaires administered periodically over a period of 24 months revealed the existence of previous episodes of low-back pain. The symptom was detected in 21% of exposed nurses but, more importantly, the presence of low-back pain at the start of the study was found to be predictive of the onset of a spinal disorder over the following months.

Healthcare workers consistently rank among the top ten jobs at high risk for work-related musculoskeletal disorders in the United States. Home healthcare workers rank the highest, together with truck drivers, while hospital nurses come after laborers but before construction workers (Collins and Menzel 2006; Hyun et al. 2011).

In 2000, government sources (US Department of Labor, Bureau of Labor Statistics) estimated an incidence of 181.6 days off work per 10,000 full-time home healthcare workers, and 90.1 days off work per 10,000 full-time hospital nurses.

Risk is present across the entire care-giving spectrum (Waters 2007), but is especially high among home healthcare workers (Howard and Adams 2010).

These findings confirm that in both the United States and Europe, back pain among healthcare workers continues to be the principal cause of morbidity in the

healthcare sector (38% of sick days attributed to LBP; Owen 2000), which is also
the cause of a high number of insurance claims and of considerable economic damage (Edlich 2005), high percentages of dropouts from healthcare professions (12%;
Stubbs et al. 1986), or requests for transfers to other duties (20%; Owen 1989). The
findings also prove how essential it is to implement global, broad-based strategies of
proven effectiveness (Fujishiro et al. 2005).

However, among healthcare workers involved in assisting and caring for disabled patients, not only the lumbosacral region is affected by disorders and diseases,
but also the upper limbs, especially the shoulders, have been shown to be involved
(Smith et al. 2003; Smedley et al. 2003; Luime 2004; Bos 2007).

Studies in the healthcare and care-giving sector have recently examined possible
correlations between stress, psychosocial factors, and musculoskeletal disorders. Feyer
(2000) found a positive correlation between low-back pain and stress; Myers, Silverstein,
and Nelson (2002) also found such a correlation between low-back and shoulder pain
with shift work and social integration. Lipscomb et al. (2002), Eriksen et al. (2006),
and Caruso and Waters (2008) found that absenteeism and low-back pain were corre-
lated with problems linked to work organization in hospitals, to shift work and extended
working hours, and home healthcare work. As mentioned previously, similar correlations were found by Smedley and colleagues in 2003 and Waters et al. in 2006.

### 2.5.2 THE ITALIAN EXPERIENCE

In Italy, epidemiological studies began to assess musculoskeletal disorders, especially of the low back, in 1994 (Larese and Fiorito). Since then many other studies
have been published (Marena et al. 1997; Baldasseroni et al. 1998, 2005; Beruffi,
Mossini, and Zamboni 1999; Ottenga et al. 2002; Squadroni and Barbini 2003; Maso
et al. 2003; Corona et al. 2004, 2005; Martinelli et al. 2004; Giannandrea et al. 2004;
Violante et al. 2004; Folletti et al. 2005; Lorusso, Bruno, and L'Abbate 2007)—all
of which evidenced a high prevalence of low-back pain among healthcare workers.
All of these studies used very different definitions of low-back pain and are thus
not very comparable. Moreover, the data concerning low-back pain have never been
compared to nonexposed controls.

Studies coordinated by the EPM (Ergonomics of Posture and Movement) Research
Unit are more interesting due to the methodologies they have employed: Risk has
consistently been assessed using the MAPO (*movimentazione e assistenza pazienti
ospedalizzati*—movement and assistance for hospitalized patients) methodology,
and low-back damage has always used acute low-back pain as an output as defined in
this chapter; lumbar disk herniation has always been confirmed instrumentally (by
computed tomography [CT] scan or magnetic resonance imaging [MRI]). Some of
these studies have been published (Battevi et al. 1999, 2006; Battevi, Menoni, and
Alvarez-Casado 2012) while others have been presented at seminars or conferences
and as such can be accessed in the relevant proceedings.

Although comparisons between the data do not take into account age, gender, and
level of exposure to the specific risk, it is interesting to consider the raw data comparison with nonexposed control groups, who may be either internal or external as
the case may be. In reading Table 2.3 it should be borne in mind that, in nonexposed

**TABLE 2.3**

**Results of Studies Coordinated by EPM Research Unit**

| Year | Sectors Assessed | No. of Units | Sample Size | % Acute Low-Back Pain in Last 12 Months | % Lumbar Disk Herniation | URL | Ref. |
|---|---|---|---|---|---|---|---|
| 1996–1999 | Hospital and geriatric wards | 216 | 3,341 | 8.4 | 7.9 | | Colombini et al. (1999) |
| 2000–2003 | Surgical blocks | 65 | 957 | 14 | 6.9 | http://www.epmresearch.org/userfiles/files/ricci%20 rischio%20blocco%20operatorio%281%29.pdf | Seminar Proceeding, Ricci, M.G. (2003) |
| 2000–2003 | Hospital and geriatric wards | 236 | 3,601 | 12.3 | — | http://www.epmresearch.org/userfiles/files/ battevi%20rischio%20PS.pdf | Seminar Proceeding, Battevi, N. (2003) |
| 2000–2003 | Emergency wards | 7 | 221 | 6 | 8 | http://www.epmresearch.org/userfiles/files/fkt%20 arezzo.pdf | |
| 2003–2007 | Physiotherapy services | 21 | 260 | 15 | 11.9 | http://www.epmresearch.org/userfiles/files/fkt%20 firenze%202.pdf http://www.epmresearch.org/userfiles/files/ Torri_rischio-danno%20fkt.pdf | 2003—Torri P. (2003); Armini F. Butticé G. (2007) |
| 2006–2008 | Hospital wards | 166 | 1,994 | 14 | 7 | http://www.epmresearch.org/userfiles/files/ seminario_5_giugno_vitelli%281%29.pdf | Seminar proceeding; Cairoli, S., Vitelli, N. (2009) |
| 2006–2008 | Outpatient clinics | 188 | 544 | 15 | 12 | http://www.epmresearch.org/userfiles/files/ seminario_5_giugno_vitelli%281%29.pdf | Seminar proceeding; Cairoli, S., Vitelli, N. (2009) |
| 2008–2009 | Geriatric wards | 31 | 411 | 6.9 | 6.8 | | Battevi et al. (2012) |
| 2013 | Surgical blocks | 43 | 723 | 8.3 | 8.3 | http://www.epmresearch.org/userfiles/ files/1-Cairoli%204%20dicembre.pdf | Seminar proceeding; Cairoli, S. (2013) |

*Note:* For details on the seminar proceedings, please access www.epmresearch.org-congressseminar.

workers, the prevalence of subjects reporting at least one episode of acute low-back pain during the previous year varies from 2.3 to 4.3, and that the lifetime prevalence of lumbar disk herniation ranges from 1% to 3% (PNLG 2005).

Although caution is advisable in reading these data, there appears to be no doubt that operators involved in manual patient-handling activities in all the sectors examined present a two to four times higher prevalence of acute low-back pain and lumbar disk herniation compared to controls, thus evidencing the possible role of specific risk.

## 2.6   CONCLUSIONS

Based on this short and admittedly not exhaustive review of the scientific literature, the following facts emerge:

- There is a definite and statistically proven correlation between the work of assisting and caring for disabled patients and increased low-back pain.
- As the work of manually handling patients increases, so does the risk of low-back pain, with a significant correlation emerging with the frequency of manual patient-handling activities; higher speed of execution may cause the tolerance level for intervertebral disk to be reached or even exceeded.
- Improper posture should be considered as an additional risk factor, along with pulling and pushing.
- The level of biomechanical overload is so high that such activities cannot be performed manually; training alone is insufficient, and the use of lifting aids is mandatory.
- There is a positive association between musculoskeletal disorders and diseases in other joints (especially the shoulder, as observed in other populations of workers exposed to manual load handling) and manually handling patient operations.
- Psychological and organizational factors must definitely be taken into account in these working populations, correlating them with low-back pain and upper limb disorders.
- Musculoskeletal disorders represent one of the main causes of morbidity in the healthcare sector, generating a high number of insurance claims and requests for transfers to other duties, and causing many workers to drop out of the profession; the significant economic and social costs involved are destined to increase going forward, unless adequate preventive policies are put in place.
- The implementation of preventive policies is possible and can be effective, but only if they are part of a specific strategic plan.

# 3 Manual Handling in the Healthcare Sector

## *The International Approach to Risk Assessment*

### 3.1 INTRODUCTION

International statistics comparable to Italian figures state that healthcare workers are among those at highest risk for musculoskeletal disorders, especially of the low back and shoulder. Therefore the problem is tackled in many countries, albeit using different approaches.

Our research group has benefited, in the effort to compare these various approaches, from the creation in 2005 of the spontaneously formed European task force named EPPHE (European Panel on Patient Handling Ergonomics) as well as from our direct participation in the drafting of ISO and CEN technical reports (TR ISO/CD 12296—"Manual Handling of People in the Healthcare Sector") published by ISO on June 1, 2012, and approved by CEN (European Committee for Standardization) in August 2013.

While the methods used to evaluate risk and preventive capabilities differ widely, there is extensive agreement over the need for an approach that begins by assessing risk and then envisages an integrated process for analyzing work organization, environment, aids, and training that, ultimately, assesses effectiveness (Hignett et al. 2014). Generally speaking, the expected results should emerge over a medium to long time frame.

Moreover, the countries in which prevention is more advanced have underscored the presence of residual risk, in spite of major efforts including financial investments, insisting that all residual risk must be tackled systematically.

In classifying the many methodological approaches to preventing manual patient-handling risk at the international level, two models have been identified that employ

a. Risk assessment tools (ranging from simple to more analytical) as the first step toward setting up preventive strategies
b. Guidelines (generally national) also encompassing risk assessment tools

The approach adopted by countries with private healthcare systems, such as the United States and Australia, commonly involves identifying risk based on patient-handling injury statistics (provided by insurance companies). This not only detects the highest risk areas but also allows the effectiveness of risk reduction strategies to be monitored economically.

## 3.2 RISK ASSESSMENT METHODS IN THE INTERNATIONAL LITERATURE

Many methods can be found in the literature for quantifying the physical risk associated with handling patients manually. These methods include:

- Laboratory studies involving the reconstruction of biomechanical overload of the lumbar spine (i.e., analysis of compressive and shear forces with relevant tolerance thresholds)
- Studies on posture during individual patient-handling movements and subsequent breakdown of the individual movement into three risk levels (i.e., task evaluation)
- Analysis of the prevalence of musculoskeletal disorders or injuries during patient-handling tasks in exposed workers

Tables 3.1 and 3.2 summarize the main characteristics of these methods and the hospital departments in which they have been or could be applied; most involve analyzing work postures and are largely based on the OWAS method (Karhu, Kansi, and Kuorinka1977). The second table lists the pros and cons of these methods.

However, for most methods, the literature fails to indicate an extremely critical aspect: the type and number of workplaces in which they were used. Interested readers may explore the abundant bibliography available on the subject.

## 3.3 GUIDELINES FOR THE PREVENTION OF RISK DUE TO PATIENT HANDLING

Internationally, guidelines represent one of the most popular tools for preventing risk due to patient-handling operations. Certain guidelines contain recommendations concerning methods (generally, checklists) for identifying and estimating risk exposure levels. As a rule, they reflect the level of awareness and sensitivity of different countries to such problems. However, while it is impossible to measure accurately the extent to which they are implemented or in how many facilities they are applied, it is clear that guidelines are employed to manage the overall risk of developing musculoskeletal disorders due to biomechanical overload.

Table 3.3 is a list of the principal guidelines at the international level; most can be accessed online.

**TABLE 3.1**

**Comparison of Different Manual Patient-Handling Assessment Methods**

| Method | Main Determinant Risk Factor/s | When and Where Applied (Also Gray Literature) | References |
|---|---|---|---|
| OWAS | Posture of all body segments, force, and frequency | Though it has not been designed for this specific goal, it has been applied in risk assessment of operating theaters. | Karhu et al. (1977) |
| LBP (low-back pain) as a function of patient lifting frequency | Lifting frequency (average frequency of manual lifting by shift) | | Stobbe et al. (1988) |
| BIPP | Full movement analysis: from preparation to implementation | | Feldstein, Vollmer, and Valanis (1990) |
| REBA | Posture of all body segments, force mainly determined by handled loads | | Hignett and McAtamney (2000) |
| PATE | Full movement analysis: from preparation to implementation | Assesses preparation to movement, caregiver's position at movement beginning and dynamic behavior | Kjellberg et al. (2000) |
| DINO | Analysis of patient transfer maneuvers | Assesses preparation, implementation, and results with 16 items; directly at workplace without movies | Johnsson et al. (2004) |
| Lift counter | Analysis of exposure to physical loads during patient care before and after the adoption of equipment Complements the use of the lift counter | Assesses exposure level to physical load, specifies the use of equipment, identifies compliance with the national guidelines, and assesses developments in the care load | Knibbe and Friele (1999) |
| MAPO | Work organization, average frequency of handling and type of patients, equipment, environment, and education and training | Considers interaction of factors | Menoni et al. (1999); Battevi et al. (2006) |

*(continued)*

**TABLE 3.1 (continued)**
**Comparison of Different Manual Patient-Handling Assessment Methods**

| Method | Main Determinant Risk Factor/s | When and Where Applied (Also Gray Literature) | References |
|---|---|---|---|
| Dortmund approach: Lumbar-overload prevention for patient-handling activities | Full movement analysis for caregiver and patient; measurement of caregiver's action forces transferred to the patient; biomechanical modeling: forces and moments at lumbar intervertebral disks | Awkward postures; exertion of high action forces; disadvantageous action-force direction; jerky movement; inadequate handling mode; disuse or misuse of aids or equipment; inadequate load-bearing capacity (e.g., due to age, gender) | Jäger et al. (2010) |
| RCN: manual handling assessments in hospitals and the community (http://www.rcn.org.uk/__data/assets/pdf_file/0008/78488/000605.pdf) | Defines three risk assessment levels: patient-based level, department or ward level, and top level; no factors quantitatively defined | Checklist provided to assess issues concerning load, posture and movement, duration frequency and job design, environment, training, organization | |

## 3.4  ISO-CEN 12296 TECHNICAL REPORT

These guidelines were drafted by the ISO/TC 159 Technical Committee's SC3 Subcommittee on Anthropometry and Biomechanics, between 2009 and 2011, and published by the ISO on June 1, 2012 (CEN ISO TR 12296 "Ergonomics: Manual Handling in the Health Care Sector").

The two main aims of the technical report (TR) are to

a. Improve the working conditions of caregivers (by reducing biomechanical overload)
b. Improve the quality of care

The technical report lays down a specific action plan for managing risk starting from risk quantification (TR Annex A), setting priorities on an integrated basis with regard to organizational aspects—for example, number of healthcare workers (TR Annex B), choice of suitable equipment/aids (TR Annex C), work environment in terms of space and passages (TR Annex D), effective staff training (TR Annex E), and ongoing monitoring of the effectiveness of the actions implemented (TR Annex F).

## TABLE 3.2
## Manual Patient-Handling Assessment Methods: Main Characteristics

| Method | Benefits | Limitations | Type of Use | References |
|---|---|---|---|---|
| OWAS | Allows scoring as well as analytical speed; considers all body segments and is useful for redesign Fits analysis of nearly all working tasks; can be used in all healthcare sectors | Analyzes posture-related aspects as the only determinant; makes it difficult to define selection criteria of postures to be analyzed; requires some time commitment | Analysis of gesture modes; can be used in an effectiveness check system | Karhu et al. (1977) |
| LBP as a function of frequency of patient lifting | Determines the manual lifting frequency and analysis speed; may predict effects on caregiver's health; can be used in hospital departments and at home | Analyzes only some types of handling (bed–wheelchair and vice versa, wheelchair–wheelchair) and action frequency is the only risk determinant considered | Rough analysis of areas/ departments more at risk | Stobbe et al. (1988) |
| BIPP | Task analysis seems to be exhaustive; seven items are used to identify a final score of movement modes through direct observation analysis; can be applied in all healthcare areas and also at home | Neglects all the other risk determinants (frequency, environment, work organization, etc.) | Can be used in an effectiveness check system | Feldstein et al. (1990) |
| REBA | Determination of scores; analysis speed useful to identify ergonomic problems associated with awkward postures and load manual handling of loads; extremely useful in hospitals and can be used in all healthcare areas | Like OWAS, practically assesses posture as the only risk determinant; actually, a load exceeding 10 kg always produces a similar score; difficult to define the selection criteria of postures to be analyzed; requires a moderate time commitment | Analysis of gesture modes; can be used in an effectiveness check system | Hignett and McAtamney (2000) |

*(continued)*

**TABLE 3.2 (continued)**
**Manual Patient-Handling Assessment Methods: Main Characteristics**

| Method | Benefits | Limitations | Type of Use | References |
|---|---|---|---|---|
| PATE | Full movement analysis: from preparation to implementation | Requires a video shot and hence may be expensive in terms of time; analyzes only manual movements and not those regarding bathrooms; neglects all other risk determinants (frequency, environment, work organization, etc.) | Can be used in an effectiveness check system | Kjellberg et al. (2000) |
| DINO | Task analysis seems to be exhaustive; final score of movement modes is identified; can be used in hospital and at home | Neglects all other risk determinants (frequency, environment, work organization, etc.) | Can be used in an effectiveness check system | Johnsson et al. (2004) |
| Lift counter | Used for monitoring purposes on a regular basis | Not specific enough for individual assessments in the patient's care plan; will require additional individual assessments | | Knibbe and Friele (1999) |
| MAPO | Allows classification into three areas (green, yellow, and red) It considers the different factors in an integrated manner and analysis of a ward needs a short evaluation time, approx. one hour (interview and inspection) | For the time being, the method was validated only for hospital wards | Applied in 400 wards for a total of approximately 6,000 exposed subjects; can be used for risk analysis in hospital wards | Menoni et al. (1999); Battevi et al. (2006) |
| Dortmund approach | Can be used for rapid evaluation of low-back loading (i.e., for identification of performance deficits) | Work-design hints are evident; provides information on the single maneuver handling but does not allow the overall risk assessment for operator or by area | | Jager et al. (2010) |

**TABLE 3.2 (continued)**
**Manual Patient-Handling Assessment Methods: Main Characteristics**

| Method | Benefits | Limitations | Type of Use | References |
|---|---|---|---|---|
| RCN:Manual handling assessments in hospitals and the community | Method can be easily used by skilled staff and is applicable in wards and communities | Since no criteria to define checklist items are available, the result of different detectors is hardly comparable; needs in-depth training for the detector and a well-structured nursing case file | Can be used for risk analysis in hospital wards and community, and also for monitoring purposes | RCN (2001) |

**TABLE 3.3**
**Principal International Guidelines**

| | |
|---|---|
| A Back Injury Prevention Guide for Health Care Providers, OSHA, California (April 2001) | http://www.dir.ca.gov/dosh/dosh_publications/backinj.pdf |
| Patient Care Ergonomics Resource Guide: Safe Patient Handling and Movement (2005) | http://www.visn8.va.gov/VISN8/PatientSafetyCenter/resguide/ErgoGuidePtOne.pdf |
| The Guide to the Handling of People: A Systems Approach, 6th ed. (2011) | http://issuu.com/backcare/docs/hop6_preview |
| Transferring People Safely—A Guide to Handling Patients, Residents and Clients in Health, Aged Care, Rehabilitation and Disability Services, 3rd ed. (July 2009) | http://www.worksafe.vic.gov.au/__data/assets/pdf_file/0011/12224/Transferring_People_Safely_-_Web.pdf |
| Dutch Guidelines for Practice (2002) | http://www.profijtvanarbobeleid.nl/html/fysieke_belasting/zorgverleners.shtml |
| AORN Guidance Statement: Safe Patient Handling and Movement in the Perioperative Setting (2007) | http://www.aorn.org |
| Patient Handling and Movement Assessments: A White Book | http://www.fgiguidelines.org/pdfs/FGI_PHAMA_whitepaper_042810.pdf |
| Guidelines for Nursing Homes. Ergonomics for the Prevention of Musculoskeletal Disorders | https://www.osha.gov/ergonomics/guidelines/nursinghome/final_nh_guidelines.pdf |

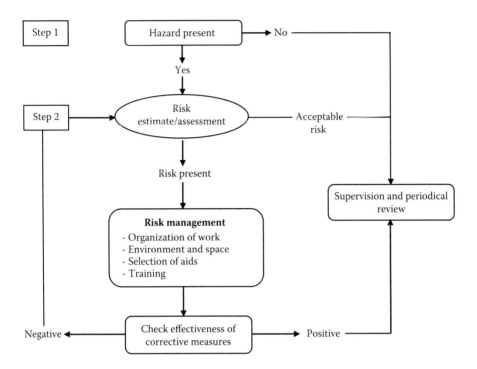

**FIGURE 3.1** Risk assessment and management: a flowchart.

Significant attention is devoted to risk assessment per the model defined by international ISO standards (11228, 11295); however, emphasis is placed not only on the need to estimate the health of workers, but also to identify problems carefully and find possible solutions. Figure 3.1 is a flowchart for risk assessment and risk management.

The MAPO (*movimentazione e assistenza pazienti ospedalizzati*—movement and assistance for hospitalized patients) method described in this volume fully complies with the scheme defined in the TR insofar as it entails a step-wise risk assessment process (i.e., hazard identification, MAPO screening, and analytical risk assessment), but also because it covers every aspect pertaining to risk management.

The task force that produced the TR discussed the emergence of a major distinction in the way risk is assessed by different cultures. While in any approach to risk assessment all that matters is clearly the result, it is nevertheless worth specifying a couple of points:

1. When identifying hazards associated with patient handling, it is essential to ask one key entry that must lead to a simple yes/no answer. The approach adopted both in the TR and the MAPO method is: "Does the service/ sector under examination involve the presence of a patient who needs to be handled?" The presence of a patient indicates that there is a hazard, which, however, is not necessarily a synonym for risk (i.e., a condition that

increases the likelihood of injury. The presence of a patient requiring handling in situations where there are adequate safe patient-handling aids, a sufficiently large number of staff per shift, enough space both for avoiding awkward postures and for using the aids, and, lastly, adequate staff training to deal with the specific risk may determine an acceptable risk level for worker health.

2. In the technical report, if a hazard is present, the next step is to perform a risk estimate/assessment. This regulation does not make a real distinction between "estimate" and "assessment" since the ISO task force opted against taking a definite stance on the issue. However, with the MAPO method it is possible to proceed in a step-wise manner, starting with a risk estimate (MAPO checklist) and, if the estimate suggests the presence of risk, going on to an analytical risk assessment (analytical MAPO). In other words, the proposed model provides the simplest and most economical approach to evaluating risk as indicated and, in fact, urged by the World Health Organization. In the following chapters, the methods will be described both for estimating risk and for carrying out an analytical assessment of risk, according to the MAPO methodology.

Starting with the step following hazard identification, the TR schematic might thus be expanded upon as follows:

- To identify any other hazards present in the healthcare sector due to biomechanical overload, the TR refers to the ISO standards, specifically ISO 12228 part 1 (lifting and carrying) and part 2 (pushing and pulling). Both of these risks may well be present in certain areas of the hospital and thus must be taken into account.
- Unlike the TR, the methodology proposed in this volume includes a key entry (i.e., "Is a hazard present or not?") for both of these aspects, so as to signal immediately the need for further evaluations.
- With regard to risks associated with lifting loads (i.e., inanimate objects such as surgical instrument trays and so on), the key entry is represented by this simple question: "Are objects weighing more than 10 kg lifted at least once daily/worker, or does the worker lift objects weighing less than this amount every 5 minutes?" If the answer is yes, then the risk assessment must be undertaken in accordance with ISO 11228-1.

The NIOSH states that 10 kg (lifted at least once per day per worker) represents the weight constant for protecting 99% of the population, both male and female.

For pushing and pulling risk, the key entry used is the following: "At least once per shift does the worker carry out a pushing and pulling operation requiring effort?" If the answer is yes, then the risk assessment must be undertaken in accordance with ISO 11228-2.

Efforts may be measured by applying the Borg scale from 0 to 10; effort in excess of 2 indicates the presence of a hazard for this type of task.

# 4 Manual Patient-Handling Risks

## *Occupational Biomechanics (Biomechanical Hazards of Manual Patient Handling)*

### 4.1 INTRODUCTION

A great deal of valuable information can be gleaned from epidemiological and risk analysis studies into the various risk factors associated with WMSDs (work-related musculoskeletal disorders). This information may contribute to a better understanding of risk and the design of ergonomically correct workstations. Various methods have been developed for measuring risk in the manual handling of loads or patients (e.g., MAPO, etc.), based on exposure to numerous risk factors.

Biomechanical studies have "measured" the degree of biomechanical overload related to individual risk factors, providing more direct and objective estimates of spinal loading (compressive forces and shear forces) and muscular fatigue levels and how the latter may lead to limiting vertebral end-plate cartilage microfractures or disk nutritional abnormalities, triggering the degeneration of the functional spinal unit.

This book aims to provide a critical, though not comprehensive, review of the main findings published in the field of occupational biomechanics, for the study of the biomechanical overload of the spine in manual patient handling, lifting/handling methods, and the assessment of patient-handling devices.

### 4.2 THE STUDY OF OCCUPATIONAL BIOMECHANICS

Occupational biomechanics as a field of study resulted from contributions (Figure 4.1) from the analysis of work motions (Taylor and Gilbreth, 1800–1900), epidemiological research into WMSDs in different work environments (Ramazzini, 1630–1700), and the mechanics of biological systems (Borelli, 1600).

Giovanni Alfonso Borelli (1608–1679), in particular, studied the human body as a machine, proving that muscular force causes a twisting moment in several muscle groups, thus paving the way for subsequent laboratory kinematics investigations based on precise mathematical models relating to mechanical physics.

In the 1600s, Bernardino Ramazzini emphasized the gradual development of musculoskeletal disorders due to the improper use of the body as a machine, especially in

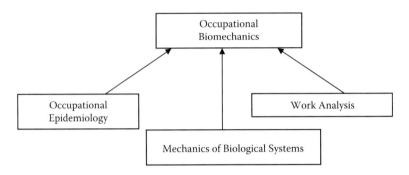

**FIGURE 4.1**    Occupational biomechanics.

the work environment. However, it took until the 1970s–1980s before epidemiological studies of this aspect of occupational medicine began to raise widespread interest and drive research into musculoskeletal disorders.

Winslow Taylor (1856–1915) and Frank Gilbreth (1868–1924) explored the effects of inefficient work organization on the human body as a machine; Gilbreth, in particular, was a pioneer of occupational ergonomics and studied the effects of poor work organization on muscle fatigue and effort.

In order to model and interpret the body's mechanical response to external physical work-related demands, over the years certain direct or in vivo measurements have been carried out. For instance, small transducers have been used to measure internal pressure on lumbar intervertebral disks in different postures and/or lifting different loads; indirect measurements have been made to determine the amount of force exerted on the internal structures of the musculoskeletal system.

Numerous in vitro studies, summarized by Chaffin et al. in 2006, have analyzed intervertebral disk compression, enabling tolerance limits to be set.

More specifically, the National Institute for Occupational Safety and Health (NIOSH) defines the "action limit" (340 kg) and "maximum limit" (650 kg) as tolerance limits for compressive forces on the low back (NIOSH 1981).

Later on, research into occupational biomechanics using *biomechanical models* defined *tolerance limits* (with respect to different forces) for the various anatomical structures of the functional spinal unit (i.e., intervertebral disks, ligaments and tendons, posterior joint processes); exceeding these limits was found to cause biological damage to the individual structures making up the spine as a whole (Marras 2008).

Between 2000 and 2008, occupational biomechanical studies shifted from primarily static models, which were nonetheless useful for estimating compressive forces acting on the intervertebral disk in a variety of work environments, to multi-dimensional models exploring the dynamics of movement; for the lumbar spine, these studies revealed the influence of physical exposure factors on low-back pain.

These studies reached the following conclusions:

- The main physical exposure factors are lifting excessive loads, asymmetrical lifting, rapid movements, awkward posture sustained by major muscular effort, and lack of coordination among the different muscle groups of the lumbar spine.

- In workplaces that commonly involve manual patient handling, these factors act synergically to increase risk levels, at times reaching and exceeding the tolerance limits of the various structures of the spine.
- Of particular importance in studying biomechanical overload of the lumbar spine is "muscle coactivation": Individual actions (such as lifting a box off a pallet and placing it on a shelf) can be broken down, at the level of the lumbar spine, into the simultaneous activation of several muscle groups (e.g., short spinal extensors, long extensors, anterior oblique, rectus abdominis, ileo psoas, etc.). Such actions result in increased biomechanical loading of the lumbar spine in the form of compressive forces, shear forces, etc. (Granata and Marras 1995).
- Only a multidimensional biomechanical assessment (i.e., kinematic and dynamic studies of the various external and internal forces involved) can quantify the biomechanical overload of the lumbar spine.

In 2006, Granata and England, analyzing estimated tolerance limits for the lumbar spine, introduced a novel concept: a breakdown of "tolerance" into the following:

- Forces/risks determining changes to the anatomical component of the functional unit of the lumbar spine
- Forces/risks determining changes to the structure of the lumbar spine as a whole

Table 4.1 shows the static tolerance limits for the various anatomical components of the lumbar spine as defined by Marras (2008) in a review of the occupational biomechanics of the lumbar spine. As shown in the table, in order to analyze individual work actions and any consequences that biomechanical overload may have on the

**TABLE 4.1**
**Possible Changes Caused by Different Applied Forces on the Lumbar Spine**

| "Target" Structures of the Spine | Type of Force | Damage to the Lumbar Spine | Certain Tolerance Limits | Remarks |
|---|---|---|---|---|
| Intervertebral disk | Compressive forces | Intervertebral disk microfractures | 520 kg (from 340 to 700 kg) | Both autopsy and biomechanical studies |
| Joint posterior | Shear forces | Posterior joint degeneration | 100 kg | As yet *uncertain* limit |
| Ligaments | Flexion | Posterior ligaments | Additional risk factor | |
| Posterior arch + disk | Extension | Anterior portion of the annulus may be damaged | Additional risk factor | |
| Posterior arch + posterior joint facet | Torsion | Degenerative changes to joint processes | Additional risk factor | |
| Intervertebral disk | Compressive forces + flexion | Microfractures to postannulus-vertebral body | From 110 to 370 kg | Possible intervertebral disk prolapse |

lumbar spine, it is necessary to consider the synergic effects of external forces that may damage the functional unit of the lumbar spine:

- Compressive forces
- Shear forces
- The possible concomitant presence of additional risk factors, such as "awkward" movements (flexion–extension or torsion), modifying the tolerance limits for certain anatomical components of the functional unit of the lumbar spine.

## 4.3  OCCUPATIONAL BIOMECHANICS STUDIES IN MANUAL PATIENT HANDLING

Many epidemiological studies undertaken over the past 20 years stress the greater frequency of lower back injuries among workers involved in patient-handling activities than in other working populations exposed to physical risk (Atlas et al. 2004).

The analysis of compressive forces ("disk loading") and tangential or shear forces during manual patient handling (Table 4.2) suggests that lifting noncooperative (NC) or completely dependent patients generates compressive forces that approach and sometimes exceed the aforesaid tolerance limits.

Even partial lifting of PC (partially cooperative) patients induces compressive forces above the action limit as defined by NIOSH studies. However, the partial lifting of fully dependent patients may reach tolerance limits, and the use of lifting devices must be evaluated if necessary and feasible (Table 4.3).

In recent years, there has been a strong focus on the use of devices to assist in patient-handling activities; in particular, lifting devices on castors are often supplied but may be seldom used. Recent studies have explored the different levels of applied force required to move a wheeled lifting device as opposed to a ceiling lift (Marras, Knapik, and Ferguson 2009).

Stress has been included in the excessive shear forces acting on the functional unit of the lumbar spine while pushing or pulling a lifter on castors; however, these are still preliminary studies that require further analysis.

The Dortmund approach should also be mentioned with reference to the biomechanical analysis of patient handling. Over the past 4 years, Jager et al. (2010) have broken down the individual tasks involved in manual patient handling into three risk levels: red, yellow, and green, depending on the compressive forces detected (Table 4.4).

Numerous variables were taken into account by all of the aforementioned studies: patients of different weight, patients more or less dependent in terms of their motor skills, comparison between different patient lifting/transfer techniques, and static single-dimensional methods versus dynamic three-dimensional methods.

The following conclusions were drawn across the board by all of the studies:

- The total lifting of fully dependent patients is far beyond the tolerance limits set for the functional units of the lumbar spine (so suitable devices must be employed).

## TABLE 4.2
## Summary of Studies into Lumbar Biomechanical Overload in Manual Patient Handling

| Author (Year) | Maneuvers Analyzed | Patient Disability (NC or PC) | Patient Weight | Result | Remarks |
|---|---|---|---|---|---|
| Dehlin and Jäderberg (1982) | Pulling up in bed Repositioning in wheelchair Turning over in bed Making bed Bed to wheelchair | NC PC | Light = <50 kg NC Light = <50 kg PC Average = 50–69 kg NC Heavy = >69 kg NC Heavy = >69 kg PC | 223 lifts analyzed Study of correlation between level of risk and several variables | The most highly risk-correlated variables are • Compressive forces • Duration of lifting • Patient disability (NC > PC) |
| Owen–Garg (1992) | Pulling up in bed Turning over in bed Wheelchair to bed Wheelchair to toilet Change sanitary pads | PC | Not reported | Compressive forces from approx. 368 to 481 kg | Correlation between Borg scale and disk load level Two-dimensional static method |
| Winkelmolen (1994) | Various techniques and maneuvers for manual lifting | PC | Not reported | All maneuvers with compressive force > 332 kg | Two-dimensional static method |
| Granata and Marras (1995) | Static analysis vs. dynamic analysis of various maneuvers | Not reported | Not reported | The static model may underestimate compressive forces by up to 45% | The static model may underestimate shear forces by up to 70% |
| Ulin (1997) | Three different manual maneuvers | NC | 56 kg; 95 kg | Obese patient with peak compressive force of approx. 1,000 kg | All manual maneuvers with compressive force > 350 kg Three-dimensional static method |

continued

**TABLE 4.2 (continued)**
**Summary of Studies into Lumbar Biomechanical Overload in Manual Patient Handling**

| Author (Year) | Maneuvers Analyzed | Patient Disability (NC or PC) | Patient Weight | Result | Remarks |
|---|---|---|---|---|---|
| Marras (1999) | Various lifting techniques<br>Liftin versus lowering<br>1 or 2 operators<br>With or without assistive devices | NC | 50 kg | Only 1 operator from 542 kg to 671 kg<br>2 operators frm 424 kg to 481 kg | Dynamic multidimensional method |
| Jang (2007) | 18 tasks defined as the most overloading | NC<br>PC | Weights lifted by a single operator (= 36–48–42–75-kg) | Compressive forces from 300 kg (PC—partial lift) to 1,300 kg (NC—full lift) | The most significant variable (force) is compressive<br>Patient lifting tasks are risky even if they only last for a few seconds |

**TABLE 4.3**

**Summary of Studies into Lumbar Biomechanical Overload in Patient Handling with Assistive Devices**

| Author (Year) | Maneuvers Analyzed | Patient Disability and Weight | Result |
|---|---|---|---|
| Ulin (1997) | Three different maneuvers with hoist | NC: 56–95 kg | Compressive force below tolerance limits |
| Marras et al. (1999) | One or two operators with lifting aids | NC: 50 kg | Compressive force below tolerance limits |
| Frigo (1999) | Maneuvers with hoist; maneuvers with minor aids | NC–PC: 56–72 kg | Compressive force below tolerance limits |
| Daynard (2001) | Pulling up in bed | NC | Compressive force from |
| | Moving from bed to stretcher with slide sheets | PC | 100 to 307 kg |
| | | Very light patient | Shear force below 100 kg |
| | | Very heavy patient | |

- Other studies (Marras et al. 1999) have shown that even when two caregivers lift relatively lightweight patients (50 kg) from bed to wheelchair or commode to chair, the compressive and/or shear forces do not drop below the tolerance limits.
- Staff involved in manual patient lifting must be trained to use lifting devices properly and to follow specific procedures.

## TABLE 4.4
## Biomechanical Assessment of Patient-Handling Activities according to the Dortmund Approach

| Activity Analyzed | Compressive Forces (Range) | Risk Level |
|---|---|---|
| Lift patient trunk from supine to sitting on bed (or vice versa) | From 180 to 540 kg | Completely cooperative PC patient<br>Correct technique + PC patient<br>In all other cases |
| From supine to sitting on the edge of the bed | From 200 to 620 kg | Correct technique + PC patient<br>In all other cases |
| Pulling patient up in bed (caregiver alongside bed) | From 210 to 810 kg | Correct technique + minor aids with PC patient<br>In all other cases |
| Pulling patient up in bed (caregiver at head of bed) | From 200 to 890 kg | Correct technique OR minor aids OR PC patient<br>In all other cases |
| Move patient toward edge of bed | From 160 to 220 kg<br>From 220 to 340 kg<br>From 330 to 580 kg | PC patient + minor aids<br>Correct technique + PC patient<br>In all other cases |
| Lift one leg of supine patient (caregiver alongside bed) | From 190 to 400 kg | In all cases |
| Lift one leg of supine patient (caregiver at foot of bed) | 180 kg | In all cases |
| Lift both legs of supine patient | From 300 to 450 kg | In all cases |
| Raise bed head with patient in bed | From 350 to 540 kg | Correct technique + fully cooperative PC patient<br>In all other cases |
| Transfer patient from sitting on edge of bed to chair | From 160 to 650 kg | Fully cooperative max. 70 kg PC patient + minor aids<br>Correct technique OR minor aids<br>In all other cases |
| Lift patient from sitting to standing or vice versa | From 190 to 310 kg<br>From 380 to 640 kg | Fully cooperative max. 70 kg PC patient + minor aids and correct technique<br>In all other cases |
| Transfer from bed to stretcher | From 230 to 240 kg | PC patient + minor aids |

### Criteria for defining risk levels

| | |
|---|---|
| Compressive forces below recommended limits (approx. 200 kg for adult females), or task considered to be acceptable for most subjects in the circumstances described | Green |
| Compressive forces within the recommended range (from 200 to 450 kg depending on age), or task considered to be acceptable in circumstances *that could be improved* | Yellow |
| Compressive forces above recommended limits for females (approx. 450 kg for young adult females), or task considered to be unacceptable for most subjects in the circumstances described | RedRed |

# 5 Methods and Criteria for Analyzing Manual Patient Handling by Healthcare Workers

## 5.1 INTRODUCTION

Developments in healthcare systems, especially in Europe, have meant that hospitals are increasingly viewed as centers for the short-term treatment of "acute" patients, while long-term care is delivered in the patient's own home or in hospices and geriatric facilities. Consequently, as far as patient-handling risk goes, there has been an absolute increase in the number of patients whose clinical condition requires them to be handled and/or lifted. Moreover, the number of nurses aged over 45 is also rising; according to recent studies (Guardini et al. 2011; Buerhaus, Staiger, and Auerbach 2000), the increasing proportion of older healthcare workers will soon make it extremely difficult to manage less than fully fit staff in areas of the healthcare system where there are patients who need to be manually handled.

The scenario is therefore destined to undergo radical change across Europe, with certain hospitals already in the forefront of this transformation, as a result of regulatory developments and improved training for nursing staff. Nursing activities will be increasingly broken down into small "modules," or sectors, with clear-cut distinctions between the tasks allocated to qualified nurses and those allocated to other healthcare professionals.

In Italian hospitals, the most common care setting is still the traditional "task-based" model.

Patient wards in both hospitals and extended care facilities are still the areas with the highest number of staff involved in patient handling and therefore have the highest number of workers exposed to the risk of biomechanical overload of the spine. A preliminary review of the most reliable literature on the subject singles out the following principal factors that, all together, characterize risk exposure in this specific setting:

- Number of dependent (or noncooperative) patients
- Type of routine total or partial patient-lifting tasks and degree of effort required
- Structural aspects of work and care environments
- Availability and actual utilization of mechanical lifting devices; training in safe patient-lifting techniques

Risk evaluations are essential for identifying which aspects can be modified to improve working conditions. Moreover, regular risk evaluations help to monitor the effectiveness of such changes. However, the methodologies employed to assess risk often require long and complicated analysis. International standards (ISO 14121) have taken this aspect into account, recommending a step-wise approach that begins with hazard identification and is followed first by risk estimation and then, if necessary, an actual risk assessment.

*In the case of manual patient handling, hazard is readily identified when the physical or mental conditions of patients are such that they cannot be moved without assistance.* Risk estimation methods are scant and woefully inadequate, although numerous methods have been proposed to evaluate risk (Hignett and McAtamney 2000; TR [technical report] 12296). The MAPO (*movimentazione e assistenza pazienti ospedalizzati*—movement and assistance for hospitalized patients) method is a risk evaluation tool that was first published in 1999 (Menoni et al. 1999; Battevi et al. 2006, 2012) by the EPM (Ergonomics of Posture and Movement) Research Unit and is widely used not only in Italy, but also in Spain (Nogareda Cuixart et al. 2011).

On the whole, the approach to risk analysis used in the MAPO method, which can also be applied to other hospital areas, replies to the following questions:

*Q:* **Who** generally needs moving or handling?
*A:* Physically dependent patients
*Q:* **What** increases the frequency of lifting tasks or biomechanical overload of the lumbar spine?
*A:* The number of healthcare workers present, mechanical lifting devices that are absent or inadequate, insufficient space or inadequate furnishings for handling patients, poor work organization, lack of training.
*Q:* **What** causes awkward postures?
*A:* Mechanical lifting devices that are absent or inadequate, insufficient space or inadequate furnishings, lack of training.

Table 5.1 lists the various risk factors defining the level of exposure to biomechanical overload due to patient-handling activities in a specific workplace. These risk factors will be more extensively described and analyzed later (MAPO index).

The information is entered into a data collection sheet, which serves two specific purposes: to calculate the MAPO risk index and to provide the information needed to draft a set of remedial actions in the specific workplace analyzed.

The risk analysis process involves two steps: an interview with the head of the ward and an on-site facility assessment. Before measuring the relevant parameters, it is advisable for the ward staff to be told what kind of information will be collected, and why. The interview with the head of the ward serves to gather information about the way the work in the ward is organized, while the ward assessment serves to gather additional details mainly relating to the equipment on hand and the environment itself, but also to check that the information gathered in the interview tallies with the situation observed in the ward.

**TABLE 5.1**
**Risk Factors for Biomechanical Overload Factors Used**
**in the MAPO Methodology**

| Risk Factor | Description |
| --- | --- |
| NC/OP | *Ratio* of average number of totally NC patients to OP over 24-hour period |
| Lifting devices factor | Ergonomic *adequacy* and *number* of devices suitable for lifting totally NC patients |
| PC/OP | *Ratio* of average number of PC patients to OP over 24-hour period |
| Minor aids factor | Ergonomic *adequacy* and *number* of devices suitable for (partially) moving PC patients |
| Wheelchairs factor | Ergonomic *adequacy* and *number* of wheelchairs |
| Environmental factor | Ergonomic *adequacy of environments* used by disabled patients for various activities |
| Training factor | *Adequacy of training* in patient-handling risk |

## 5.2 ORGANIZATIONAL ANALYSIS

The purpose of this initial stage is to assess the "care load" in terms of patient handling and largely to estimate the frequency with which an operator performs patient-handling activities. To overcome the problem of choosing which information to collect and how to collect it, a special data collection sheet has been drawn up (Figure 5.1).* Figure 5.2 shows how to collect useful data and define the main risk factors.

This information should be clearly conveyed during the interview to avoid misunderstandings about the terminology used. It is worth noting that risk exposure is not measured on a one-off basis, but rather over a period of 1 year, and it refers to *average patient-handling activities* performed over a 24-hour period.

Besides identifying certain specific aspects of the ward, such as its name and the number of beds, the interviewer will start by counting the average number of operators working over three shifts. It is worth stressing that the term "handling" refers to the total or partial lifting and transferring of patients either manually or using mechanical devices (MPH).

### 5.2.1 PATIENT-HANDLING OPERATORS

The ward staff involved in patient-handling activities and thus potentially exposed to MPH risk is broken down into job categories, described, and quantified in order to assess whether the shifts are properly designed. In other words, the total number of staff members must be greater than the number of operators assigned to the three shifts. The interviewer must check whether there are enough attendants to allow for the necessary contractual days off.

---

* *Note to the Reader*: Figures that are forms appear at the end of each chapter.

The next step will be to define how many of the operators performing MPH activities are working on each shift (morning, afternoon, and night) and if they generally handle patients alone or in pairs (the number of pairs of operators carrying out MPH activities per shift is also reported). The sum total of operators assigned to patient-handling tasks is used to calculate the operators present (OP) factor, taking into account the following details.

*Number of operators present* is the sum total of operators assigned to patient-handling tasks present in the morning, afternoon, and night (note that this is *not* the total number of ward staff).

Most operators these days still work over three shifts, but alternative work schedules are becoming increasingly common. These include part-time and "vertical" (i.e., nurses working two or three shifts per week) or "horizontal" (i.e., nurses working 4 hours every morning, Monday through Friday).

Table 5.2 provides an example of data collected for a ward and its staff. From the practical standpoint, as indicated in the checklist, if operators are present for part of a full shift, they must be calculated as unit fractions of the number of hours worked in the shift. The section of the sheet indicating this aspect is "no. part-time operators." There is also space in the data sheet to indicate the start and end of each operator's working hours and then for entering the number of hours worked versus the total number of hours in the shift.

If the work schedule envisages only the rotation of three shifts, the MPH operators should be viewed as *homogeneous* because it can be assumed that over the course of a year, there will be no major differences in the number and nature of shifts worked. In this case, the risk index calculated for the workplace will also correspond to the risk index calculated for the individual workers. In other cases, the ward risk index will be measured to identify the elements required to put in place a risk reduction plan.

## 5.2.2 Description and Number of Dependent Patients

The concept of "dependent" may pose several terminological difficulties, insofar as it covers a variety of aspects (i.e., cognitive, functional, mobility related, etc.) and can be measured using different tools (such as the Barthel index, for example) now commonly employed in long-term and elderly care facilities. The interviewee must understand that for the purposes of detecting MPH-related risk, the term "dependent" is used to define patients who need ward staff to "help" them move or be moved fully or partly, regardless of their pathology. Therefore, it is an aspect of dependency that relates both to the patient's mobility function and to the ward's specific organization and environment.

Another critical element is the number of dependent patients. This requires calculating the average number of dependent patients present in the ward. To facilitate this calculation, there is a section of the data sheet that lists the clinical condition of the patient, as well as the patient's degree of dependency. Dependent patients may be defined as *totally noncooperative* (NC) or *partially cooperative* (PC), according to their need to be lifted/moved totally or partially.

## TABLE 5.2
## Example 1: Data Concerning Total Nursing Staff Engaged in MPH

In the cardiology ward with 26 beds, 23 operators are engaged in MPH: 18 nurses and five nursing aids. Usually, they work over three shifts: five nurses on the morning shift (from 7.00 a.m. to 2.00 p.m.), three nurses on the afternoon shift (from 2.00 p.m. to 9.00 p.m.), and three nurses on the night shift (from 9.00 p.m. to 7.00 a.m.). Other operators are present during the morning (from 8.00 a.m. to 12.00 a.m.).

| Description of the Healthcare Facility | | |
|---|---|---|
| Hospital: | Ward: cardiology | Ward code: |
| No. beds: 26 | Average hospital stay (days): | Date: |

| Total Nursing Staff Engaged In Patient Handling: Mark the Total Number of Operators/Category | | |
|---|---|---|
| Nursing staff: 18 | Nurses aides: 5 | Other: |

**No. Operators Engaged in MPH over Three Shifts: Indicate the Number of Operators on Duty per Shift**

| Shift | Morning | Afternoon | Night |
|---|---|---|---|
| Shift schedule: (00:00 to 00:00) | From 7:00 to 2:00 | From 2:00 to 9:00 | From 9:00 to 7:00 |
| No. operators over entire shift | 4 | 3 | 2 |
| | | (A) Total operators over entire shift = | 9 |

**No. Part-Time Operators: Indicate the Exact Number of Hours Worked and Calculate Them as Unit Fractions (in Relation to the Overall Duration of the Shift)**

| No. Part-Time Operators Present | Hours Worked in Shift: 00:00 to 00:00 | Unit Fraction | Unit Fraction By No. Operators |
|---|---|---|---|
| | From 8.00 a.m. to 12.00 p.m. | 4/7 | 0.57 |
| | From__to___ | | |
| | From__to___ | | |
| (B) Total operators (as unit fractions) present by shift duration = | | | **0.57** |
| Total no. of operators engaged in MPH over 24 hours (Op): add the total number of operators present over the entire shift (A) to the total number of part-time operators (B) = | | 9.57 | Op |

**Totally noncooperative** means that the patient is unable to assist in being moved and needs to be completely lifted.

**Partially cooperative** means that the patient may be able to assist in being moved and only needs to be partially lifted.

Besides being logical, biomechanical studies (Marras 2008; Jager et al. 2007; Waters, Nelson, and Proctor 2008) have also proven that biomechanical overload

**TABLE 5.3**

**Example 2: Data Collection Sheet for Types of Patients**

The cardiology ward has an average of twelve dependent patients: eight NC (including five elderly patients with multiple comorbidities and three with various neuropathies) and four extremely elderly cardiovascular patients.

Types of Patients

Totally noncooperative patients (NC) are patients who *need to be fully lifted* in transfer/repositioning operations. Partially cooperative patients (PC) are patients who need *only partial lifting*.

Disabled patients (D): 12 (indicate average number per day)

Noncooperative patients (NC): 8. Partially cooperative patients (PC): 4

| Disabled Patients | No. NC | No. PC |
|---|---|---|
| Elderly with multiple concomitant diseases | 5 | |
| Hemiplegic | | |
| Surgical | | |
| Severe stroke | | |
| Dementia | | |
| Other neurologic diseases | 3 | |
| Fracture | | |
| Bariatric | | |
| Other | | 4 |
| Total | 8 | 4 |

of the lumbar spine is closely linked to the type of handling activities performed, hence the need for this distinction.

Once the purpose of the assessment has been explained, the section entitled "types of patients" may be completed; Table 5.3 provides an example.

However, it is worth emphasizing that, generally speaking, the head nurse or "senior" operator is able to identify these aspects with great accuracy: the paired *t*-test reported by the authors in the literature (Menoni et al. 1999) proved that there were no differences between the objective and subjective findings, provided the interviewee had actually worked in the ward for a certain length of time.

Should the interviewee be unable to provide this information, another tool can be used to detect objectively the average number of NC and PC patients in the ward; in this case, it is advisable to carry out without delay a daily assessment of the types of patients in the ward and transfer the data to the relevant checklist (Figure 5.3) so as to indicate accurately the degree of dependency of each patient. If the number of dependent patients changes during the course of the week, the assessment must be performed for at least 7 consecutive days.

At the end of the assessment, the average number of totally NC and PC patients present during the observation week is specified. This number, depicted as a simple mathematical formula at the bottom of the data sheet, will be reported in the MAPO form (Figure 5.3).

The medium long-term solution undoubtedly lies in creating a database for these details, which can be obtained from nursing care records; in this way, the hospital IT system can provide the real-time information needed to quantify risk.

### 5.2.3 DESCRIPTION OF OPERATOR TRAINING

It should be noted that the term "training" refers both to the process of educating (i.e., transmitting knowledge and skills) and delivering training properly (i.e., teaching how to use equipment). In order to identify these aspects indirectly, with reference to patient-handling risk, the following information must be obtained during the interview:

- **Type** of training delivered
- **Duration** of training course
- **Number** of operators currently on the staff involved in training
- **Time elapsed** between training course and current risk evaluation process
- **Assessment** of training effectiveness

The type and duration of training define the concept of training adequacy; training must address the specific risk at hand and illustrate the epidemiological aspects relating to the number of exposed workers and to the prevalence and incidence of work-related musculoskeletal disorders (WMSDs) among healthcare workers. Biomechanics must also be touched on, to stress lumbar disk loading in relation to weights lifted and awkward lifting postures. The training course should, of course, include practice sessions focusing on the proper use of mechanical devices for the total and partial handling of dependent patients. A theoretical and practical course of this kind should last **at least** 6 hours.

It is therefore not enough simply to distribute pamphlets on patient handling or to ask the sales reps of equipment suppliers to explain how to use their devices. Neither of these approaches provides a complete picture of the issues at hand, nor do they feature the necessary educational component.

The number of operators involved is also of the utmost importance: In light of possible staff turnover, the ratio of trained to untrained staff should be 3:1. In other words, at least 75% of patient-handling staff should be adequately trained.

The last two aspects concerning training (i.e., time elapsed since the last training course and an assessment of the effectiveness of training) analyze the training strategy put in place in the specific care facility under examination. Since training is an educational *process,* the adoption of periodic monitoring demonstrates that management is genuinely determined to take steps to modify staff behavior. With regard to this aspect, in Italy and elsewhere there seems to be a greater focus on assessing the effectiveness of training using various reliable tools, rather than on the methods employed for delivering the training.

However, since this information is not always available to our "expert" interviewee, it may be necessary to contact the head of nursing or the hospital training

**TABLE 5.4**

**Education and Training in Data Collection Sheet**

| Operator Education and Training | | | |
|---|---|---|---|
| Training | | Information | |
| Theoretical/practical course delivered | ☐ Yes  ☐ No | Training only on how to use equipment | ☐ Yes  ☐ No |
| If yes, how many months ago?  Months ____ | | Only provided brochures on MPH | ☐ Yes  ☐ No |
| How many hours/operator?  Hours ____ | | | |
| If yes, how many operators? | | If yes, how many operators? | |
| Was effectiveness measured and documented in writing? | | | ☐ Yes  ☐ No |

center. In other cases, where a quality control system is in place, the information may be obtained directly in the wards.

The section of the risk detection sheet relating to this aspect is shown in Table 5.4.

### 5.2.4 DEFINITION OF PATIENT-HANDLING TASKS

Another section of the checklist describes the patient-handling tasks routinely undertaken in the ward, with a distinction made between tasks requiring total patient lifting (TL) and tasks requiring partial moving or lifting of the patient (PL) (Table 5.5). The way wards are organized should make it easier to complete this part of the checklist, since the various care tasks, including patient handling, are usually scheduled and standardized according to a set timetable. The checklist includes different sections for manual patient handling with and without the use of lifting aids.

The purpose of this description is twofold: if there are plans to purchase lifting devices, this section helps to decide which ones best suit the needs of each specific ward, but it will also help to calculate objectively the percentage of patient-handling operations requiring mechanical devices.

To make it easier to collect the data, routine patient-handling operations in the ward have been broken down into specific tasks and specific shifts. By "routine," we mean handling operations that are performed on a daily basis. Conceptually, handling operations may be grouped together as follows:

- **Lifting/turning a patient in bed** (turning from back to side, correcting posture of bedridden patient, pulling up in bed)
- **Transferring a patient from bed** (bed to wheelchair and vice versa; bed to stretcher and vice versa)
- **Moving a seated patient** (lifting from sitting to upright and vice versa; to wheelchair and toilet and vice versa)

"Other" refers to any other manual patient-handling activities not previously listed.

Turning patients in bed should be noted only if *a part of the patient's body needs lifting* (since part of the patient's body has to be lifted, it is always classified as PL

**TABLE 5.5**

**Example 3: Data Collection on Types of Tasks Currently Carried Out**

In the cardiology ward there are generally eight NC patients and seven PC patients at any given time. Most patient-handling operations involve providing bedside assistance to the disabled patient: Three tasks involving pulling NC patients up in bed are performed during morning shift and two on the afternoon shift, while PC patients are pulled up in bed three times in the morning, three times in the afternoon, and twice during the night shift. Patients are turned around in bed or their posture changed three times during the morning and afternoon shifts, and twice during the night shift.

| Patient-Handling Tasks Currently Carried Out In One Shift | | | | | | |
|---|---|---|---|---|---|---|
| **Manual Handling:** Describe routine tasks involving total or partial patient lifting. Indicate the *number of tasks* per shift involving manual patient handling | **Total Lifting (TL) Without Equipment** | | | **Partial Lifting (PL) Without Equipment** | | |
| | Morning | Afternoon | Night | Morning | Afternoon | Night |
| | A | B | C | D | E | F |
| Pulling up in bed | xxx | xx | | xxx | xxx | xx |
| Turning over in bed (to change position) | | | | xxx | xxx | xx |
| Bed to wheelchair and vice versa | | | | | | |
| Lifting from seated to upright position | | | | | | |
| Bed to stretcher and vice versa | | | | | | |
| Wheelchair to toilet and vice versa | | | | | | |
| Other | | | | | | |
| Other | | | | | | |
| Total: calculate the total for each column | 3 | 2 | 0 | 6 | 6 | 4 |
| Number of total or partial manual lifting tasks | A + B + C = TL | | 5 | D + E + F = PL | | 16 |

even if the patient is completely noncooperative). Of course, only partially cooperative patients can be lifted from sitting to upright, and as such do not require total lifting.

This section of the data collection sheet is designed to roughly calculate the total number of care tasks that require patient handling (both total and partial lifting). The sum total of columns A, B, and C corresponds to the total number of tasks requiring TL, while the sum total of columns D, E, and F corresponds to the total number of handling operations requiring the partial lifting/moving of the patient. Note that only manual handling tasks must be reported in this section (Table 5.6). (Table 5.7, relating to patient-handling tasks using adequate lifting aids, follows the same rationale.)

**TABLE 5.6**

**Manual Handling**

| Patient-Handling Tasks Currently Carried Out In One Shift | | | | | | |
|---|---|---|---|---|---|---|
| **Manual Handling:** Describe routine tasks involving total or partial patient lifting Indicate the *number of tasks* per shift involving manual patient handling | **Total Lifting (TL) Without Equipment** | | | **Partial Lifting (PL) Without Equipment** | | |
| | **Morning** | **Afternoon** | **Night** | **Morning** | **Afternoon** | **Night** |
| | A | B | C | D | E | F |
| Pulling up in bed | | | | | | |
| Turning over in bed (to change position) | | | | | | |
| Bed to wheelchair and vice versa | | | | | | |
| Lifting from seated to upright position | | | | | | |
| Bed to stretcher and vice versa | | | | | | |
| Wheelchair to toilet and vice versa | | | | | | |
| Other | | | | | | |
| Other | | | | | | |
| Total: calculate the total for each column | | | | | | |
| Number of total or partial manual lifting tasks | A + B + C = TL | | | D + E + F = PL | | |

Accordingly, the sum total of columns G, H, and I corresponds to the number of total patient-lifting tasks using lifting aids (ATL), while the sum total of columns L, M, and N corresponds to the number of partial patient-lifting tasks using lifting aids (APL). This makes it extremely simple to compare the total number of aided patient-lifting operations (over three shifts) with the total number of routine handling operations, either total or partial. It is thus possible to determine the percentage of total versus partial patient-handling operations using lifting aids:

Percent of aided total lifting operations (% ATL): ATL/(TL + ATL)

Percent of aided partial lifting operations (% APL): APL/(PL + APL)

It is worth pointing out that this part of the checklist is not based only on the interviewee's own subjective experience; the following section on lifting aids present in wards where manual patient-handling operations are carried out, along with informal interviews with staff on duty during the next visit to the ward, enables

**TABLE 5.7**
**Aided Handling**

| Aided Handling:<br>Describe routine tasks involving total or partial patient lifting using available equipment<br>Indicate the *number of tasks* per shift involving aided patient handling | Total Lifting (Aided) | | | Partial Lifting (Aided) | | |
|---|---|---|---|---|---|---|
| | Morning | Afternoon | Night | Morning | Afternoon | Night |
| | G | H | I | L | M | N |
| Pulling up in bed | | | | | | |
| Turning over in bed (to change position) | | | | | | |
| Bed to wheelchair and vice versa | | | | | | |
| Lifting from seated to upright position | | | | | | |
| Bed to stretcher and vice versa | | | | | | |
| Wheelchair to toilet and vice versa | | | | | | |
| Other | | | | | | |
| Other | | | | | | |
| Total: calculate the total for each column | | | | | | |
| Aided handling total (ATL) or partial (APL) lifting | G + H + I = ATL | | | L + M + N = APL | | |
| Percentage of aided total lifting operations (% ATL) | ATL/(TL + ATL) | | | | | |
| Percentage of aided partial lifting operations (% APL) | | | | APL/(PL + APL) | | |

the accuracy of the interviewee's replies to be verified. Although the information is reported via an interview, this specific section of the data sheet is included on the page describing the lifting equipment present in the ward.

## 5.3   ANALYSIS OF PATIENT-HANDLING DEVICES

The analysis of patient-handling devices during the ward visit is of particular importance, requiring the interviewer to be very well trained. To begin, the methodologies for conducting such an analysis are scarce, and, as well, the term "equipment" is used to describe anything that decreases or eliminates biomechanical overload, especially of the lumbar spine, while handling dependent patients. Equipment may thus include devices for handling totally and partially cooperative patients, as well as devices for transporting them.

**TABLE 5.8**
**Patient-Lifting Devices**

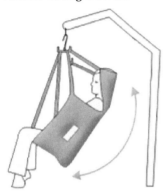
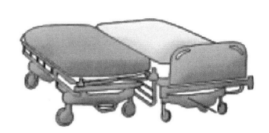

As evidenced in Table 5.8, the checklist considers patient-lifting devices (hoists), stretchers, ergonomic beds, and aids described as "minor." To facilitate filling in the data sheet, the section describing patient beds comes after the section describing patient rooms. In any event, beds qualify as equipment if they have certain ergonomic characteristics, can raise the patient's upper body, and generally help to shift the patient from sitting to standing and prevent him or her from slipping toward the foot of the bed.

The equipment must be counted and classified by type of device and ergonomic characteristics. Minor aids are defined as devices that can reduce the number of certain partial patient-handling operations as well as decrease biomechanical overload of the lumbar spine.

Besides common minor aids, such as slide mats or sheets, this section will also describe active lifting devices, or standing hoists, to help lift partially cooperative patients from sitting to standing; their use requires a certain amount of cooperation on the part of the patient.

The interviewer will essentially analyze the adequacy of the equipment with respect to the various patient-handling tasks indicated previously. Moreover, any inadequacies with respect to the environment or the patients for whom the equipment is routinely used should also be reported.

We refer the reader to Chapter 10 for other specific equipment analysis criteria.

### 5.3.1 DESCRIPTION OF WHEELCHAIRS

The last equipment to be considered is wheelchairs. Generally speaking, wheelchairs should

- Allow for the use of a wheeled hoist
- Help to transfer partially cooperative patients
- Feature a compact design

From the ergonomic standpoint, several characteristics have been considered that, if lacking, may either increase the frequency of manual lifting or cause biomechanical overload of the lumbar spine.

The wheelchair analysis consists of

- Counting the total number of wheelchairs assigned to (and actually used in) the ward
- Grouping the wheelchairs by type (using the letters A, B, C, etc.)
- Assessing any ergonomic inadequacies with respect to each type of wheelchair (i.e., lack of ergonomic requirements)
- Calculating the scores for each column (i.e., the sum total for each column) and multiplying them by the number of wheelchairs of a specific type
- Adding up the column scores to obtain the total wheelchair score
- Calculating the mean inadequacy score for wheelchairs (MSWh) by dividing the total wheelchair score by the total number of wheelchairs in the ward

Please note the following definition of a "cumbersome backrest" (with a score of 1 = ergonomically poor):

1. Backrest total thickness over 6 cm, or
2. Backrest total height (from upper edge to floor) over 90 cm, or
3. Backrest leaning back more than 10°

To assign a maximum width score of 1, the width of the wheelchair must be over 70 cm, generally measured to the outside of the rear wheels. Table 5.9 shows an example of the section on wheelchairs. Wheelchairs belonging to the same brand but with different characteristics must be described separately, as if they are different types.

---

**TABLE 5.9**

**Example 4: Types of Wheelchairs**

There are eight wheelchairs in the ward, two of which are new and ergonomically acceptable, while three have nonextractable armrests and a cumbersome back, and three do not brake efficiently.

| Wheelchairs | | Type of Wheelchair (No.) | | | | | |
|---|---|---|---|---|---|---|---|
| | | A | B | C | D | E | F |
| **Wheelchair Features and Inadequacy Score** | **Score** | 2 | 3 | 3 | — | — | — |
| Poor maintenance | | | | | | | |
| Malfunctioning brakes | 1 | | | X | | | |
| Nonremovable armrest | 1 | | X | | | | |
| Nonremovable footrest | | | | | | | |
| Cumbersome backrest | 1 | | X | | | | |
| Width exceeding 70 cm | 1 | cm | cm | cm | cm | cm | cm |
| Column score (no. wheelchairs × sum of scores) | | | 6 | 3 | | | |

*Note:* Mean wheelchair score (MSWh) = total wheelchair score/no. wheelchairs; |9/8 = 1.12| MSWh.

---

This part of the checklist is compiled following the same principles as the environmental assessment; it includes the ergonomic aspects that together determine the score, as well as actual descriptions of the equipment itself. For example, poor brakes resulting from inadequate maintenance will be important for the purposes of patient handling, while poor maintenance—consisting, for instance, of a torn backrest—does not impair efficient patient handling.

Clearly, the method for calculating the mean inadequacy score for wheelchairs leads to a maximum value of 4; therefore, a score of 4 represents the maximum degree of wheelchair inadequacy.

## 5.4   ENVIRONMENTAL ASSESSMENT

In order to assess manual patient-handling risk, attention must also be paid to the environments in which patients are lifted or transferred. These include

1. Bathrooms
2. Toilets
3. Patient rooms

Each specific environment must be accurately described and, for wheelchairs, there is a general description as well as a score for ergonomic inadequacies. Ergonomic inadequacy is defined as the inability to use equipment (e.g., hoists, wheelchairs, etc.), awkward postures for attendants approaching patients, and the inability to leverage a partially cooperative patient's residual motor skills. This part of the on-site assessment requires the use of a simple tape measure, the aim being to *determine the mean ergonomic inadequacy score for the ward environment.*

Most of the characteristics to be analyzed also depend on the degree of compliance with national regulations governing architectural barriers.

### 5.4.1   BATHROOMS

This section of the data sheet (Figure 5.1) adopts the same approach as for wheelchairs. Bathrooms may be centralized or en suite, and each will be described in separate sections, but they are grouped by type and together generate a mean inadequacy score for bathrooms (MBS). The calculations employ the same procedure as for wheelchairs/commodes.

One potential challenge may lie in defining "inadequate space for using aids." In the literature, there are some recommendations in this respect that might help (e.g., there are a few recommendations in this respect that might help—e.g., ISO TR 12296).

The UK National Health Service considers adequate maneuvering space to be a minimum turning circle with a diameter that will vary depending on certain variables:

1. For maneuvering the wheelchair of an independent user, a clear radius of 168 cm
2. For maneuvering the wheelchair of a nonindependent user, a clear radius of at least 215 cm
3. For turning and using a wheeled patient hoist, a diameter of at least 225 cm

In other cases, recommendations are provided concerning circulation and communication spaces—however, without indicating whether the space is adequate for the use of patient-handling aids. For instance, a space suitable for using lifting devices might be entirely unsuitable for the posture of attendants.

From the practical standpoint, besides relying on such guidance, it is essential for the interviewer to "try out" the aids in the ward and draw the relevant conclusions. Understandably, patient-handling devices may not need to be rotated fully and may be moved using a variety of maneuvers.

The effective clear width of bathroom and bedroom doors is a hot topic. The literature includes specifications such as the health building note, which advises a door width of 115 cm for wheelchairs and wheeled devices; bed and patient trolley movement recommends an opening of 174 cm. Leaf and a half doors are in common use to ensure suitable access. However, it is advisable as a good technical rule for doors to have an effective clear width of at least 85 cm. Most of the equipment routinely used in wards (e.g., wheelchairs, hoists on castors, trolleys, and so on) can readily be moved without additional turning maneuvers with this door width.

A common finding is the presence of nonremovable obstacles, such as a shower tray that is not on the same level as the floor. This must be reported and assigned the appropriate inadequacy score.

## 5.4.2 Bathrooms and Toilets

As before (Table 5.10), this section is for calculating the mean inadequacy score of toilets (MSWC); however, a few ergonomic aspects must be taken into special consideration. In describing these facilities, the interviewer must specify whether toilets are en suite or in a centralized bathroom. With regard to "free space insufficient to turn wheelchair," reference is made to regulations governing the removal of architectural barriers.

Nevertheless, the MAPO recommendations are as follows:

1. Space beside WC for positioning wheelchair: at least 80 cm
2. Grab handles (with height specifications)
3. Height of WC: at least 45–50 cm

The space to position a wheelchair beside the WC may be on one side only. Grab handles on each side of the WC should be removable so as to avoid constituting a lateral obstacle.

## 5.4.3 Patient Rooms

Over recent years, stricter regulations have given rise to significant changes in patient room design. Eight- or ten-bed wards have generally become a thing of the past, and most patient rooms now comfortably accommodate just one or two beds. Countless tasks and operations are performed in patient rooms, including patient handling, that call for a certain number of requirements to be met. The checklist gathers the following critical information for patient-handling purposes (Table 5.11).

**TABLE 5.10**
**Types of Bathrooms: Toilets/WCs**

| Toilets (WCs) | | Type of Toilet (WC) | | | | | | | |
|---|---|---|---|---|---|---|---|---|---|
| Features and Inadequacy Score | Score | En suite | En suite | En suite | Centralized Bathroom | | | | |
| | | No. | No. | No. | No. | No. | No. | No. | |
| Free space insufficient to turn around wheelchair | 2 | | | | | | | | Total no. toilets (WCs) \|___\| |
| Door opening inward (not outward) | | | | | | | | | |
| Insufficient height of WC (below 50 cm) | 1 | | | | | | | | |
| WC without grab bars | 1 | | | | | | | | |
| Door width less than 85 cm | 1 | | | | | | | | |
| Space at side of WC less than 80 cm | 1 | | | | | | | | Total WC score: |
| Column score (no. toilets × sum of scores) | | | | | | | | | |

*Notes:* If grab bars are present but inadequate, indicate reason for inadequacy in remarks and count as absent. Mean WC score (MSWC) = total WC score/no. WCs. \|___\| MSWC.

The method for calculating the mean room ergonomic inadequacy score (MRS) is the same as the one used for assessing other environmental aspects and wheelchairs. However, the characteristics to be analyzed are different: for instance, the free space between beds, between the beds and the wall, and between the foot of the bed and the wall. These distances must be measured to ensure that there is sufficient space to use various types of patient-handling devices. Recommendations regarding these aspects have appeared in the literature since the late 1800s, generally specifying the space requirements for hospital beds. The range goes from a minimum of 6.91 up to 13.3 square meters, based on the type of maneuvers to be performed at the patient's bedside. Commonly, an area of 9 square meters is indicated, which the data sheet defines as sufficient (90 cm between beds and 120 cm at the foot of the bed); this corresponds almost exactly with the range indicated by the American Institute of Architects in 2006.

Patient beds also need to have enough space underneath to position a patient hoist on castors. A space of 15 cm is generally sufficient, but the interviewer needs to check whether there are mechanisms below the bed for regulating its height. Also

**TABLE 5.11**

**Patient Room Configuration**

| Rooms: Features and Inadequacy Score | Score | Patient Rooms | | | | | |
|---|---|---|---|---|---|---|---|
| | | No. Rooms | No. Rooms | No. Rooms | No. Rooms | No. Rooms | |
| Number of beds per room | | | | | | | |
| Space between beds or between bed and wall less than 90 cm | 2 | | | | | | Total no. rooms |
| Space between foot of bed and wall less than 120 cm | 2 | | | | | | \|____\| |
| Presence of nonremovable obstacles | | | | | | | |
| Fixed beds with height less than 70 cm | | cm | cm | cm | cm | cm | |
| Unsuitable bed that needs to be partially lifted | 1 | | | | | | |
| Inadequate side flaps | | | | | | | |
| Door width | | cm | cm | cm | cm | cm | |
| Space between bed and floor less than 15 cm | 2 | cm | cm | cm | cm | cm | |
| Beds with two wheels or no wheels | | | | | | | Total room score: |
| Height of armchair seat less than 50 cm | 0,5 | | | | | | |
| Column score (no. rooms × sum of scores) | | | | | | | |

*Note:* Mean room score (MSR) = total ward score/total no. rooms \|____\| MSR.

common are beds where the head of the bed can be manually raised, potentially straining the spine of the operator, especially when the patient is lying in the bed.

Lastly, there may be a problem with totally or partially noncooperative patients using excessively low armchairs; operators may end up adopting awkward postures to move the patient from sitting to standing and vice versa.

### 5.4.4 DESCRIPTION OF PATIENT BEDS

As stated before, even if beds are regarded to all intents and purposes as equipment, the section describing beds in the data sheet comes after the section on patient rooms, for purely practical reasons. At this point, after the visit, the interviewer will be able to describe the types of beds present analytically, as seen in Table 5.12.

**TABLE 5.12**
**Height-Adjustable Beds**

| Height-Adjustable Beds | | | | | | |
|---|---|---|---|---|---|---|
| Description of Beds | | Nr | Electric adjustable | Mechanical adjustable | Nr of sections | Manual lifting of bed head or foot |
| BED A: | | | Yes No | Yes No | 1 2 3 | Yes No |
| BED B: | | | Yes No | Yes No | 1 2 3 | Yes No |
| BED C: | | | Yes No | Yes No | 1 2 3 | Yes No |
| BED D: | | | Yes No | Yes No | 1 2 3 | Yes No |

## 5.5 CRITERIA FOR DEFINING THE SPECIFIC MAPO INDEX OF WARDS

Before going into how the MAPO index is calculated, it is worth remembering how the index was initially created and later validated by association studies linking the MAPO index with low-back injury (see Chapter 6). In 1994 the EPM Research Unit began a series of observational studies on activities performed in hospital wards; the studies involved the description, via diary entries, of all patient-handling operations, awkward attendant positions, and structural and environmental aspects and equipment.

By analyzing these descriptions in around 200 wards, it was possible to define the aspects that all together either increase the frequency of patient-lifting actions or increase biomechanical overload of the lumbar spine. This preliminary effort led to a hypothesis for the quantitative definition of the MAPO exposure index, based on the following formula:

$$(NC/OP \times LF) + (PC/OP \times AF) \times WF \times EF \times TF$$

where

NC and PC represent the number of noncooperative and partially cooperative patients
OP is the number of operators working over three shifts
LF is the lifting devices factor
AF is the minor aids factor
WF is the wheelchairs factor
EF is the environmental factor
TF is the training factor

To facilitate calculating the index, a legend is provided together with a summary of the values of the individual factors (legend linked to checklist in Figure 5.2).

By analyzing each individual factor, the formula tells us that certain parameters can be obtained directly from the checklist. These include the number of operators working over three shifts and the number of noncooperative patients, broken down into totally NC and PC.

It is worth reiterating that OP indicates the total number of operators performing manual patient-handling activities over three shifts. NC and PC indicate the *average number* of noncooperative and partially cooperative patients present in the ward.

The other risk factors for which values have been established beforehand must be drawn from the information contained in the data sheet:

1. LF (lifting devices factor)
2. AF (minor aids factor)
3. WF (wheelchairs factor)
4. EF (environmental factor)
5. TF (training factor)

### 5.5.1   VALUE OF THE LIFTING DEVICES FACTOR (LF)

It should be noted that this factor comes into play in wards where there are totally noncooperative patients, who therefore need to be fully lifted in transfer operations. The use of the term "lifting devices factor" should not be misunderstood as the need to include *only* lifting devices as equipment for hoisting up patients. This risk factor in fact refers to "any devices that can be used to fully lift up a patient."

The following scores may be assigned to this factor:

$$\text{Absent or inadequate + insufficient} = 4$$

$$\text{Present but inadequate or insufficient} = 2$$

$$\text{Adequate + sufficient} = 0.5$$

Clearly, two parameters must be used to assign the correct score to this factor: *adequacy and numerical sufficiency*.

For the main lifting devices, numerical **sufficiency** is defined as

- At least one patient-lifting device for every eight fully noncooperative patients, or
- At least one height-adjustable stretcher (where patients are generally transferred to and from flat surfaces) for every eight totally noncooperative patients, or
- A height-adjustable *three-segment* bed for every patient in the unit

While other publications also give a definition of numerical sufficiency, requirements may differ based on the way in which the unit is organized, the workplace environment as a whole, the degree of operator training, and how many transfers calling for total patient lifting are carried out using equipment.

The situation is **adequate** when at least *90% of total patient lifting operations are performed using lifting aids.* There are practical reasons for this cut-off level: Certain lifting operations (such as lifting a patient off the ground) may be carried out only occasionally; they do not, however, change the routine way in which attendants use the equipment. These data are found in the checklist.

$$\text{Percent of total aided lifting operations} = ATL/(TL + ATL)$$

*When lifting is not aided, the equipment is reported as* **absent (LF = 4).**

## 5.5.2   VALUE OF THE MINOR AIDS FACTOR (AF)

This factor refers only to partially cooperative patients, and in this case two criteria apply: adequacy and numerical sufficiency. The possible scores are

$$\text{Absent or (inadequate + insufficient)} = 1$$

$$\text{Present and adequate + sufficient} = 0.5$$

Here, too, **adequacy** is defined as *90% of operations to lift or handle noncooperative patients performed using aids.* These data are found in the checklist:

$$\text{Percent of } \textbf{total aided lifting operations} = APL/(PL + APL)$$

Numerical sufficiency is defined as indicated:

- Presence of sliding sheets or sliding boards + two other minor aids, or
- Presence of sliding sheets or sliding boards + 100% of ergonomic beds (height-adjustable three-section beds)

Here, too, numerical **sufficiency** requirements may differ based on the way the unit is organized, the workplace environment as a whole, the degree of operator training, and how many transfers calling for partial patient lifting are carried out using equipment.

## 5.5.3   VALUE OF THE WHEELCHAIR FACTOR (WF)

It is thus possible to determine the mean ergonomic inadequacy score for wheelchairs (MSWh); the maximum score is 4. In this case, numerical sufficiency is defined as "the presence of at least 50% as many wheelchairs as there are dependent patients" in the ward (i.e., the sum of NC and PC patients). Table 5.13 indicates how to score this factor correctly.

   The maximum ergonomic inadequacy score (4) has been divided into three groups: low, medium, and high, and these criteria are also used to assign a score to an environmental factor. It should be stressed that this factor is calculated for *all* dependent patients.

**TABLE 5.13**
**Value of Wheelchair Factor (WF)**

| Inadequacy | Low | | Medium | | High | |
|---|---|---|---|---|---|---|
| Mean wheelchair score (MSWh) | 0–1.33 | | 1.34–2.66 | | 2.67–4 | |
| Numerical sufficiency | No | Yes | No | Yes | No | Yes |
| Value of WF | 1 | 0.75 | 1.5 | 1.12 | 2 | 1.5 |

**TABLE 5.14**
**Value of Environmental Factor (EF)**

| Inadequacy | Low | Medium | High |
|---|---|---|---|
| Environmental mean score (MSE) | 0–5.8 | 5.9–11.6 | 11.7–17.5 |
| Environmental factor score (EF) | 0.75 | 1.25 | 1.5 |

### 5.5.4 VALUE OF THE ENVIRONMENTAL FACTOR (EF)

In the data collection sheet there is a description of the ergonomic inadequacy of the workplace in which patients are handled: bathrooms and toilets (WCs) and patient rooms. A mean ergonomic inadequacy score is calculated for each of these environments as follows:

MSB: mean inadequacy score for bathrooms
MSWC: mean inadequacy score for toilets
MSR: mean inadequacy score for patient rooms

In order to identify the score for this factor, it is necessary to calculate the sum of all three inadequacy scores (mean environment score = MSB + MSWC + MSR = MSENV) and check the corresponding EF score in Table 5.14.

The environment mean score is calculated directly on the checklist and, as indicated previously for wheelchairs, the score is determined based on the ergonomic inadequacy class into which it falls (low, medium, or high, obtained by dividing the maximum potential environmental inadequacy score into three groups).

As for the wheelchair factor, the environmental factor does not make distinctions between NC and PC patients, since it affects all dependent patients.

### 5.5.5 VALUE OF THE TRAINING FACTOR (TF)

Mention has already been made of training as a risk factor whose characteristics are reported during the interview (i.e., type of training and whether tested for effectiveness). An **adequate training course** is defined as a theoretical/practical course of at least 6 hours with some of the practical training devoted to the use of lifting aids.

**TABLE 5.15**
**Value of Training Factor (TF)**

| Characteristics Observed | TF Value |
|---|---|
| Adequate **training course** held no more than 2 years before the risk assessment, attended by 75% of ward staff engaged in patient handling | 0.75 |
| Training course held more than 2 years before the risk assessment, attended by 75% of ward staff (engaged in patient handling) and tested for effectiveness | 0.75 |
| Adequate **training course** held no more than 2 years before the risk assessment, attended by between 50% and 75% of ward staff (engaged in patient handling). | 1 |
| Only delivery of information (or specific brochures) to 90% of ward staff, followed by effectiveness testing | 1 |
| *Training not delivered or not compliant with aforementioned conditions* | 2 |

**TABLE 5.16**
**Value of MAPO Exposure Level**

| MAPO Index | Exposure Level |
|---|---|
| 0 | Absent (white) |
| 0.1–1.5 | Negligible (green) |
| 1.51–5 | Medium (yellow) |
| >5 | High (red) |

Once the necessary information has been collected, the training factor is scored as shown in Table 5.15. Effectiveness testing must be demanded; therefore, during the interview, the documentation attesting objectively to training delivery must be acquired.

In Figure 5.4, summarizing the MAPO risk factors identifies the exposure level shown in Table 5.16.

## 5.6  CHARACTERISTICS OF ITALIAN EXTENDED CARE FACILITIES

With Italy's aging population and a tendency for hospitals to focus on treating acute patients, there has been a major increase in the number of extended care facilities; consequently, the number of operators exposed to specific risk has also risen. While hospital stays are becoming shorter and shorter, residential aged care facilities are being utilized for correspondingly longer periods. This has entailed a number of developments, some favorable to risk assessment and others less so. As well, certain risk factors need to be interpreted slightly differently in a nursing home compared to a hospital ward. For the purposes of this chapter, it is worth listing these differences:

1. The number of noncooperative patients in an extended care facility is generally very high.
2. In Italy, patient-handling operations are carried out by workers belonging to two professional categories: OTA (**o**peratori **t**ecnici **a**ddetti all'**a**ssistenza)

or nursing aides, rather than by qualified nurses, of whom there is a chronic shortage.

3. Staff do not always work three shifts. This is not a serious issue since the checklist also allows for "unit fractions" of staff present in the shift; however, the final MAPO risk index score may fail to represent accurately the risk to which operators performing manual patient-handling activities are exposed. Accordingly, it may be necessary to reconstruct the percentage of aided patient-handling activities carried out per shift, a number that can easily be obtained from the MAPO data sheet. The overall MAPO index for the ward in which these operators work will be more valuable for preventive purposes rather than assigning a specific risk exposure level.

4. It is easier to describe patient-handling activities on a per-shift basis, since the organization of the workload is the result of a specific planning process and the activities can readily be summarized by the head of the ward. In the specific checklist, the aforementioned patient-handling operations include additional tasks carried out in the ward—for example, "repositioning in wheelchair" or "changing diapers" (if the patient needs to be partially or fully moved).

Therefore, the risk assessment checklist for such facilities is slightly different compared to those described elsewhere in this chapter.

The calculation of the MAPO risk index is the same as that described in the previous sections, with the exception of the numerical sufficiency factor for wheelchairs; in this case there must be enough wheelchairs for 80% of the noncooperative patients present, as opposed to 50% as indicated previously.

## 5.7 MAPO RISK ESTIMATION AND SCREENING

Risk is estimated here using the MAPO screening process (Figure 5.5). This screening tool is lean and simple, and well suited to evaluating the presence and level of risk in hospital and care facilities. It should not be used as an alternative to the more complex MAPO analytical assessment, but rather as a filter for deciding whether or not a more detailed analysis is required.

In order to use the checklist properly, staff must be trained to use the MAPO method. Although the checklist is seemingly simpler, it still requires adequate theoretical and practical training to ensure fully reliable results. At the facility level, adequately trained technical staff may use the checklist to screen all the jobs involving manual patient handling, solely via interviews, and using the same predetermined environmental factor and wheelchairs factor. Scores are assigned to the LF (lifting devices), AF (minor aids factor), and TF (training) factors following the same rules as for the analytical procedure, while the EF and WF are assigned different scores depending on *the type of facility being analyzed (i.e., either a hospital or residential care facility).*

The final calculation of the risk exposure score is based on the formulae in Tables 5.17 and 5.18. In both cases, the scores are the result of experience acquired by our team since the 1990s. In the facilities there is no shortage of nonergonomic situations that increase the number of handling operations required to manage patients, and that frequently hamper the proper use of lifting aids. This is why a conservatively

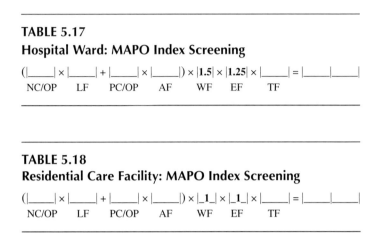

**TABLE 5.17**
**Hospital Ward: MAPO Index Screening**

$$(|\underline{\quad}| \times |\underline{\quad}| + |\underline{\quad}| \times |\underline{\quad}|) \times |\mathbf{1.5}| \times |\mathbf{1.25}| \times |\underline{\quad}| = |\underline{\quad}|\underline{\quad}|$$
$$\quad\text{NC/OP}\quad\;\text{LF}\qquad\text{PC/OP}\qquad\text{AF}\qquad\text{WF}\qquad\text{EF}\qquad\text{TF}$$

**TABLE 5.18**
**Residential Care Facility: MAPO Index Screening**

$$(|\underline{\quad}| \times |\underline{\quad}| + |\underline{\quad}| \times |\underline{\quad}|) \times |\mathbf{1}| \times |\mathbf{1}| \times |\underline{\quad}| = |\underline{\quad}|\underline{\quad}|$$
$$\quad\text{NC/OP}\quad\;\text{LF}\qquad\text{PC/OP}\qquad\text{AF}\qquad\text{WF}\qquad\text{EF}\qquad\text{TF}$$

intermediate score has been used for the factor envisaged by the analytical methodology. The same applies to the wheelchairs factor: In many hospitals there is often a shortage of wheelchairs and a lack of proper maintenance.

In the case of residential aged/elderly care facilities, given the types of patients involved and the existence of specific regulations, there is usually sufficient space for patient-handling operations and a wheelchair for each patient; therefore, the score assigned is neutral and does not automatically raise the risk exposure level.

Data can thus be quickly collected so as to classify risk levels in different areas as "absent," "negligible," "medium," or "high" and to see how such areas are distributed within the specific facility. The analysis can thus form the basis for planning the necessary corrective actions and investments.

## 5.8  WARD CHARACTERISTICS WITH RESPECT TO DIFFERENT SHIFT SCHEDULES

Developments, for better or worse, in hospital care in Europe and internationally have led to different organizational scenarios for exposed staff:

1. Rotation over three shifts of all workers exposed to patient-handling risk, with average shifts of 8 hours
2. Rotation over three shifts and simultaneous presence of staff members always working day shifts and/or part-time shifts
3. No shift rotation and staff members exposed to risk (in the ward) for 6 hours
4. Shifts of 12 hours with or without rotation

For the first two scenarios, the MAPO index not only identifies the risk level for the whole ward, but also determines the risk exposure of the workers involved in patient-handling tasks (except for part-time workers or staff who only work day shifts). Conversely, a few points need to be emphasized with respect to scenarios 3 and 4:

- The aim of the MAPO method is to assess the level of risk associated with manual patient handling in wards *featuring the organizational and environmental characteristics described here.* With this approach it is possible to set *priorities for corrective action* and define preventive strategies for reducing risk levels.
- Accordingly, even in situations such as those described under scenarios 3 and 4, the MAPO index remains the same and essentially defines the risk level in the ward in question.
- Wards organized as described in scenarios 3 and 4 leave a number of factors that are difficult to analyze (e.g., type of handling tasks per individual patient, frequency of handling tasks per individual attendant, etc.).

There are two ways to calculate manual patient-handling risk exposure levels in staff working 6- or 12-hour shifts (essentially based on the length of the shift versus the average 8-hour shift). Take, for instance, the MAPO index calculated for the cardiology ward in Chapter 12 (example 1): a cardiology ward with the same number of operators, but organized over four 6-hour shifts. In this case, to assess the exposure of the operator working 6 hours, the MAPO index is reduced by a quarter (i.e., by 2 hours for an average 8-hour shift):

$$\text{MAPO index for 6-hour shifts} \rightarrow 2.47 \times 0.75 = 1.85$$

Similarly, for 12-hour shifts, the MAPO index will be multiplied by 1.5:

$$\text{MAPO index for 12-hour shifts} \rightarrow 2.47 \times 1.5 = 3.7$$

### DATA COLLECTION SHEET – RISK ASSESSMENT FOR MANUAL PATIENT HANDLING IN WARDS

### 1. INTERVIEW

| DESCRIPTION OF THE HEALTHCARE FACILITY | | |
|---|---|---|
| HOSPITAL: | WARD: | WARD CODE: |
| Nr BEDS: | AVERAGE HOSPITAL STAY (days): | DATE: |

**Nr OF OPERATORS ENGAGED IN MPH:** indicate the total number of operators per job category

| Nursing staff: | Nurses aides: | Other: |
|---|---|---|

**Nr of OPERATORS ENGAGED IN MPH OVER 3 SHIFTS:** indicate the number of operators on duty per shift

| SHIFT | morning | Afternoon | night |
|---|---|---|---|
| Shift schedule: (00:00 to 00:00) | from_____to_____ | from_____to_____ | from_____to_____ |
| Nr of operators over entire shift | | | |

(A) Total operators over entire shift =

**Nr of OF PART-TIME OPERATORS:** indicate the exact number of hours worked and calculate them as unit fractions (in relation to the overall duration of the shift).

| Nr of part-time operators present | Hours worked in shift: (00:00 to 00:00) | Unit fraction | (unit fraction by Nr of operators) |
|---|---|---|---|
| | from_____to_____ | | |
| | from_____to_____ | | |
| | from_____to_____ | | |

(B) Total operators (as unit fractions) present by shift duration =

**TOTAL Nr OF OPERATORS ENGAGED IN MPH OVER 24 HOURS (Op):** add the total number of operators present over the entire shift (A) to the total number of part-time operators (B)       **OP**

Is the work carried out by two nurses? If it is, indicate the number of 2-nurse teams per shift:

1° morning _____ 2° afternoon_____ 3° night _____

| TYPE OF PATIENTS: |
|---|

"Totally Non-Cooperative" patients (**NC**) are patients who <u>need to be fully lifted</u> in transfer/repositioning operations.

"Partially Cooperative" patients (**PC**) are patients who <u>need only partial lifting.</u>

DISABLED PATIENTS (**D**)_____ (indicate average number per day)
Non-Cooperative patients (**NC**) Nr_____ Partially Cooperative patients (**PC**) Nr _____

| DISABLED PATIENTS | Nr NC | Nr PC |
|---|---|---|
| Elderly with multiple concomitant diseases | | |
| Hemiplegic | | |
| Surgical | | |
| Severe stroke | | |
| Dementia | | |
| Other neurologic diseases | | |
| Fracture | | |
| Bariatric | | |
| Other | | |
| Total | | |

**FIGURE 5.1**   Ward sheet. *(continued)*

FONDAZIONE IRCCS CA' GRANDA
OSPEDALE MAGGIORE POLICLINICO
CLINICA DEL LAVORO – MILAN (ITALY)
ERGONOMICS SECTION

| OPERATOR EDUCATION AND TRAINING | | | | | |
|---|---|---|---|---|---|
| **EDUCATION AND TRAINING** | | | **INFORMATION** | | |
| Attended theoretical/practical course | ☐ YES | ☐ NO | Training only on how to use equipment | ☐ YES | ☐ NO |
| if YES, how many months ago? and how many hours/operator | Months _____ hours _____ | | Only provided brochures on MPH | ☐ YES | ☐ NO |
| if YES, how many operators? | | | if YES, how many operators? | | |
| Was EFFECTIVENESS measured and documented in writing? | | | ☐ YES | ☐ NO | |

| PATIENT HANDLING TASKS CURRENTLY CARRIED OUT IN ONE SHIFT: | | | | | | |
|---|---|---|---|---|---|---|
| **MANUAL HANDLING:** describe routine tasks involving total or partial patient lifting | **Total lifting (TL) WITHOUT EQUIPMENT** | | | **Partial Lifting (PL) WITHOUT EQUIPMENT** | | |
| indicate the **number of tasks** per shift involving manual patient handling | morning | afternoon | night | morning | afternoon | night |
| | A | B | C | D | E | F |
| ☐ pulling up in bed | ☐☐☐☐ | ☐☐☐☐ | ☐☐☐☐ | ☐☐☐☐ | ☐☐☐☐ | ☐☐☐☐ |
| ☐ turning over in bed (to change position) | | | | ☐☐☐☐☐ | ☐☐☐☐☐ | ☐☐☐☐☐ |
| ☐ bed-to-wheelchair and viceversa | ☐☐ | ☐☐ | ☐☐ | ☐☐ | ☐☐ | ☐☐ |
| ☐ lifting from seated to upright position | | | | ☐☐ | ☐☐ | ☐☐ |
| ☐ bed-to-stretcher and viceversa | ☐☐ | ☐☐ | ☐☐ | ☐☐ | ☐☐ | ☐☐ |
| ☐ wheelchair-to-toilet and viceversa | ☐☐ | ☐☐ | ☐☐ | ☐☐ | ☐☐ | ☐☐ |
| ☐ other | ☐☐ | ☐☐ | ☐☐ | ☐☐ | ☐☐ | ☐☐ |
| ☐ other | ☐☐ | ☐☐ | ☐☐ | ☐☐ | ☐☐ | ☐☐ |
| **TOTAL:** calculate the total for each column | | | | | | |
| **Number of total (TL) or partial (PL) manual lifting tasks** | A+B+C = TL | | | D+E+F=PL | | |

| AIDED HANDLING: describe routine tasks involving total or partial patients lifting using available equipment | **Total lifting (TL) AIDED** | | | **Partial Lifting (PL) AIDED** | | |
|---|---|---|---|---|---|---|
| Indicate the **number of tasks** per shift involving aided patient handling | morning | afternoon | night | morning | afternoon | night |
| | G | H | I | L | M | N |
| ☐ pulling up in bed | ☐☐☐☐ | ☐☐☐☐ | ☐☐☐☐ | ☐☐☐☐ | ☐☐☐☐ | ☐☐☐☐ |
| ☐ turning over in bed (to change position) | | | | ☐☐☐☐☐ | ☐☐☐☐☐ | ☐☐☐☐☐ |
| ☐ bed-to-wheelchair and viceversa | ☐☐ | ☐☐ | ☐☐ | ☐☐ | ☐☐ | ☐☐ |
| ☐ lifting from seated to upright position | | | | ☐☐ | ☐☐ | ☐☐ |
| ☐ bed-to-stretcher and viceversa | ☐☐ | ☐☐ | ☐☐ | ☐☐ | ☐☐ | ☐☐ |
| ☐ wheelchair-to-toilet and viceversa | ☐☐ | ☐☐ | ☐☐ | ☐☐ | ☐☐ | ☐☐ |
| ☐ other | ☐☐ | ☐☐ | ☐☐ | ☐☐ | ☐☐ | ☐☐ |
| ☐ other | ☐☐ | ☐☐ | ☐☐ | ☐☐ | ☐☐ | ☐☐ |
| **TOTAL:** calculate the total for each column | | | | | | |
| **AIDED handling total (ATL) or partial (APL) lifting** | G+H+I = ATL | | | L+M+N=APL | | |
| **% OF AIDED TOTAL LIFTING OPERATIONS (% ATL)** | $\frac{ATL}{(TL + ATL)}$ | | | | | |
| **% OF AIDED PARTIAL LIFTING OPERATIONS (% APL)** | | | | $\frac{APL}{(PL + APL)}$ | | |

**FIGURE 5.1** *(continued)*  Ward sheet. *(continued)*

FONDAZIONE IRCCS CA' GRANDA
OSPEDALE MAGGIORE POLICLINICO
CLINICA DEL LAVORO – MILAN (ITALY)
ERGONOMICS SECTION

## 2.ON SITE INSPECTION

**EQUIPMENT FOR DISABLED PATIENT LIFTING/TRANSFER \***

| EQUIPMENT DESCRIPTION | | Nr | Lack of essential requirements | Lack of adaptability to patients or environment | Lack of maintenance |
|---|---|---|---|---|---|
| LIFTING EQUIPMENT type: | | | YES  NO | YES  NO | YES  NO |
| LIFTING EQUIPMENT type: | | | YES  NO | YES  NO | YES  NO |
| LIFTING EQUIPMENT type: | | | YES  NO | YES  NO | YES  NO |
| Adjustable STRETCHER type: | | | YES  NO | YES  NO | YES  NO |
| Adjustable STRETCHER type : | | | YES  NO | YES  NO | YES  NO |

**OTHER AIDS (MINOR AIDS):**

| EQUIPMENT DESCRIPTION | | Nr | Lack of essential requirements | Lack of adaptability to patients or environment | Lack of maintenance |
|---|---|---|---|---|---|
| SLIDING SHEETS | | | YES  NO | YES  NO | YES  NO |
| STANDING HOISTS type: | | | YES  NO | YES  NO | YES  NO |
| ERGONOMIC BELTS: | | | YES  NO | YES  NO | YES  NO |
| SLIDING BOARDS: | | | YES  NO | YES  NO | YES  NO |
| OTHER: | | | YES  NO | YES  NO | YES  NO |

\* N.B. : Attach floor plan to assess available space for more equipment and if there is an equipment storage room

| WHEELCHAIRS: | Score | Type of wheelchair | | | | | | Total Nr of wheelchairs |
|---|---|---|---|---|---|---|---|---|
| WHEELCHAIR FEATURES AND INADEQUACY SCORE | | A Nr | B Nr | C Nr | D Nr | E Nr | F Nr | |_____| |
| Poor maintenance | | | | | | | | |
| Malfunctioning brakes | 1 | | | | | | | |
| Non-removable armrest | 1 | | | | | | | |
| Non-removable footrest | | | | | | | | |
| Cumbersome backrest | 1 | | | | | | | Total wheel chairs score: |
| Width exceeding 70 cm | 1 | Cm | Cm | Cm | Cm | Cm | Cm | |
| **Column score** (Nr of wheelchairs x sum of scores) | | | | | | | | |

MEAN WHEELCHAIRS SCORE (MSWh) = Total wheelchair score / Nr of wheelchairs   = |_____|

**FIGURE 5.1** *(continued)*   Ward sheet. *(continued)*

FONDAZIONE IRCCS CA' GRANDA
OSPEDALE MAGGIORE POLICLINICO
CLINICA DEL LAVORO – MILAN (ITALY)
ERGONOMICS SECTION

epm

**STRUCTURAL FEATURES OF ENVIRONMENT BATHROOMS (centralized or individual in rooms)**

**TYPES OF BATHROOMS WITH SHOWER/BATH:**

| | Score | TYPE OF BATHROOM WITH SHOWER/BATH | | | | | | | |
|---|---|---|---|---|---|---|---|---|---|
| | | En-suite | | | Centralized bathrooms | | | | |
| BATHROOMS WITH SHOWER/BATH: FEATURES AND INADEQUACY SCORE | | Nr | Nr | Nr | Nr | Nr | Nr | Nr | Total Nr of bathrooms \|__\| |
| Free space inadequate for use of aids | 2 | | | | | | | | |
| Door opening inwards (not outwards) | | | | | | | | | |
| No shower | | | | | | | | | |
| No bath | | | | | | | | | |
| Door width less than 85 cm | 1 | cm | cm | cm | cm | cm | cm | cm | Total bathroom score |
| Non-removable obstacles | 1 | | | | | | | | |
| **Column score (Nr bathrooms x sum of scores)** | | | | | | | | | |

Mean bathroom score (**MBS**) = Total bathroom score/total Nr bathrooms : |______| **MBS**

TOILETS (WC):

| | Score | TYPE OF TOILETS (WC) | | | | | | | |
|---|---|---|---|---|---|---|---|---|---|
| | | En-suite | En-suite | En-suite | Centralized bathrooms | | | | |
| TOILETS: FEATURES AND INADEQUACY SCORE | | Nr | Nr | Nr | Nr | Nr | Nr | Nr | Total Nr of toilets (WC) \|__\| |
| Free space insufficient to turn around wheelchair | 2 | | | | | | | | |
| Door opening inwards (not outwards) | | | | | | | | | |
| Insufficient height of WC (below 50 cm) | 1 | | | | | | | | |
| WC without grab bars* | 1 | | | | | | | | |
| Door width less than 85 cm | 1 | | | | | | | | |
| Space at side of WC less than 80 cm | 1 | | | | | | | | Total WC score: |
| **Column score (Nr toilets x sum of scores)** | | | | | | | | | |

* if GRAB BARS are present but inadequate, indicate reason for inadequacy in notes and count as absent

Mean WC score **(MSWC)** = total WC score/Nr WCs: |______| **MSWC**

**NOTES**

______________________________________________

______________________________________________

______________________________________________

**FIGURE 5.1** *(continued)* Ward sheet. *(continued)*

FONDAZIONE IRCCS CA' GRANDA
OSPEDALE MAGGIORE POLICLINICO
CLINICA DEL LAVORO – MILAN ITALY
ERGONOMICS SECTION

| PATIENT ROOM CONFIGURATION | Score | PATIENT ROOMS | | | | | |
|---|---|---|---|---|---|---|---|
| ROOMS: FEATURES AND INADEQUACY SCORE | | Nr of rooms | Nr of rooms | Nr of rooms | Nr of rooms | Nr of rooms | |
| Number of beds per room | | | | | | | |
| Space between beds or between bed and wall less than 90 cm | 2 | | | | | | Total Nr of rooms \|___\| |
| Space between foot of bed and wall less than 120 cm | 2 | | | | | | |
| Presence of non-removable obstacles | | | | | | | |
| Fixed beds with height less than 70 cm | | Cm Nr | Cm Nr | Cm Nr | Cm Nr | Cm Nr | |
| Unsuitable bed that needs to be partially lifted | 1 | | | | | | |
| Inadequate side flaps | | | | | | | |
| Door width | | Cm | cm | cm | cm | cm | |
| Space between bed and floor less than 15 cm | 2 | cm | cm | cm | cm | cm | |
| Beds with 2 wheels or no wheels | | | | | | | |
| Height of armchair seat less than 50 cm | 0,5 | | | | | | Total room score: |
| **Column score (Nr of rooms × sum of scores)** | | | | | | | |

Mean room score **(MSR)** = total ward score /total Nr rooms |_____| **MSR**

INDICATE IF BATHROOMS (OR WHEELCHAIRS) ARE NOT USED BY DISABLED PATIENTS (CONFINED TO BED)
☐ YES   ☐ NO

**MEAN ENVIRONMENT SCORE: MSB + MSWC + MSR = |____| MSENV**

| HEIGHT-ADJUSTABLE BEDS | | | | | | |
|---|---|---|---|---|---|---|
| DESCRIPTION OF BEDS | | Nr | Electric adjustable | Mechanical adjustable | Nr of sections | Manual lifting of bed head or foot |
| BED A: | | | YES   NO | YES   NO | 1   2   3   4 | YES   NO |
| BED B: | | | YES   NO | YES   NO | 1   2   3   4 | YES   NO |
| BED C: | | | YES   NO | YES   NO | 1   2   3   4 | YES   NO |
| BED D: | | | YES   NO | YES   NO | 1   2   3   4 | YES   NO |

**FIGURE 5.1** *(continued)*   Ward sheet.

FONDAZIONE IRCCS CA' GRANDA
OSPEDALE MAGGIORE POLICLINICO
CLINICA DEL LAVORO – MILAN (ITALY)
ERGONOMICS SECTION

### WARD MAPO: HOW TO ASSIGN "VALUES" TO RISK FACTORS

**Total number of Operators engaged in MPH over 24 hours (OP):** indicate the sum total patient handling operators present in the morning, afternoon and night. If operators are present for a portion of the shift, calculate them as unit fractions in relation to the number of hours worked in the shift.

**Disabled patients (D):** indicate the average number of NC and PC patients routinely present in the ward. Fully non-cooperative (**NC**) refers to patients who need to be completely lifted when they have to be transferred. Partially cooperative (**PC**) refers to patients who only need partial lifting.

**VALUE OF LIFTING DEVICES FACTOR (LF):** "Lifting aids" refers to all the devices and aids that can be used to fully lift patients. Active lifters with chest loops are not considered as "lifting factors" but rather as "minor aids" factors.

**DETAILED DESCRIPTION OF NUMERICAL SUFFICIENCY FOR "LIFTING AIDS FACTOR":**
-       at least 1 **patient lifting device** for every 8 fully non-cooperative patients;
                                                                       or
-       at least one **height-adjustable stretcher** (where patients are generally transferred to and from flat surfaces) for every 8 fully non–cooperative patients;
                                                                       or
-       a **height-adjustable three-segment bed** for every patient in the unit.

The term **adequate** refers to equipment that meets the needs of the ward **i.e.**
**When at least 90% of operations to totally lift patients are performed using lifting aids**
**When no operations to lift patients are performed using lifting aids then the definition to be given is "lifting aids absent". (THEREFORE: LF =4)**

Below are the scores for the "LIFTING DEVICES FACTOR":

**absent** or ( **inadequate + insufficient)   = 4**
**present** but **inadequate or insufficient = 2**
**adequate + sufficient                = 0.5**
Indicate the percentage of total patient lifts using lifting aids = ATL/ (TL + ATL) = |__|__|__|

**VALUE OF MINOR AIDS FACTOR (AF):**

**absent** or **inadequate or insufficient = 1**
**adequate and sufficient = 0.5**
Indicate the percentage of partial patient lifts using lifting aids = APL/ (PL + APL) = |__|__|__|
**Adequate and sufficient "minor aids":**
Here too, when at least 90% of operations to partially lift patients are performed using aids.
There must be:
- a sliding sheet or sliding board + two of the other minor aids indicated
                                             OR
- a sliding sheet or sliding board + ergonomic bed (100% of beds in the ward)

**VALUE OF WHEELCHAIRS FACTOR (WF):**
To define the wheelchairs factor score, it is necessary to calculate the MEAN INADEQUACY SCORE obtained in the data collection sheet (**MSWh**) in relation to the numerical sufficiency of the wheelchairs as indicated in the table below:

| WHEELCHAIRS FACTOR (WF) | | | | | | |
|---|---|---|---|---|---|---|
| Mean score observed (**MSWh**) | 0 – 1.33 | | 1.34 – 2.66 | | 2.67 – 4 | |
| Numerical sufficiency | NO | YES | NO | YES | NO | YES |
| WF SCORE | 1 | 0.75 | 1.5 | 1.12 | 2 | 1.5 |

**FIGURE 5.2**   Ward sheet's legend. *(continued)*

**Numerical sufficiency of wheelchairs** – numerical sufficiency refers to situations where the number of wheelchairs is more than one wheelchair per 50% of disabled patients (**D**). For hospital geriatric wards, more than 80% of disabled patients (also in elderly care facilities or residential care homes).

| VALUE OF ENVIRONMENT FACTOR (EF): |
|---|

To assign the environment factor, the "mean inadequacy score" for the environment (**MSENV**) calculated on the data collection sheet is divided into three equidistant ranges (i.e. <u>low, medium and high inadequacy)</u>, as shown in the figure below

| **MSENV** | 0-5.8 | 5.9-11.6 | 11.7-17.5 |
|---|---|---|---|
| **VALUE OF ENVIRONMENT FACTOR** | 0.75 | 1.25 | 1.5 |

| VALUE OF THE TRAINING FACTOR (TF): |
|---|

| Characteristics observed | TF value |
|---|---|
| Adequate **training course** held no more than 2 years before the risk assessment, attended by 75% of ward staff engaged in patient handling | 0.75 |
| Training course held more than 2 years before the risk assessment, attended by 75% of ward staff (engaged in patient handling), and tested for effectiveness | 0.75 |
| Adequate **training course** held no more than 2 years before the risk assessment, attended by between 50% and 75% of ward staff (engaged in patient handling) | 1 |
| Only explanations (or specific brochures) given to 90% of ward staff (engaged in patient handling), followed by effectiveness testing | 1 |
| TRAINING NOT DELIVERED OR NOT COMPLIANT WITH AFORE MENTIONED CONDITIONS | 2 |

An **adequate training course** is defined as a theoretical/practical course of at least 6 hours with some of the practical training devoted to the use of lifting aids.

**FIGURE 5.2** *(continued)*    Ward sheet's legend.

FONDAZIONE IRCCS CA' GRANDA
Ospedale Maggiore Policlinico
CLINICA DEL LAVORO – MILAN -ITALY
ERGONOMICS SECTION

**SHEET FOR TYPE OF PATIENTS IN WARDS**

HOSPITAL :_____ WARD :_____ ward code : _____

N. beds: _____ date _____

**TYPE OF PATIENTS:**
By totally Non-Cooperative patients (**NC**) we mean the patient who <u>is to be fully lifted</u> in transfer operations. By Partially Collaborating patient (**PC**) we mean the patient who is <u>only partially lifted.</u> (**A** = <u>patient self sufficient</u> )

<u>Indicate for every bed the type of patient's disability</u>

| beds | monday | | | tuesday | | | wednesday | | | thursday | | | friday | | | saturday | | | sunday | | |
|---|---|---|---|---|---|---|---|---|---|---|---|---|---|---|---|---|---|---|---|---|---|
| | A | NC | PC | A | NC | PC | A | NC | PC | A | NC | PC | A | NC | PC | A | NC | PC | A | NC | PC |
| **1** | | | | | | | | | | | | | | | | | | | | | |
| **2** | | | | | | | | | | | | | | | | | | | | | |
| **3** | | | | | | | | | | | | | | | | | | | | | |
| **4** | | | | | | | | | | | | | | | | | | | | | |
| **5** | | | | | | | | | | | | | | | | | | | | | |
| **6** | | | | | | | | | | | | | | | | | | | | | |
| **7** | | | | | | | | | | | | | | | | | | | | | |
| **8** | | | | | | | | | | | | | | | | | | | | | |
| **9** | | | | | | | | | | | | | | | | | | | | | |
| **10** | | | | | | | | | | | | | | | | | | | | | |
| **11** | | | | | | | | | | | | | | | | | | | | | |
| **12** | | | | | | | | | | | | | | | | | | | | | |
| **13** | | | | | | | | | | | | | | | | | | | | | |
| **14** | | | | | | | | | | | | | | | | | | | | | |
| **15** | | | | | | | | | | | | | | | | | | | | | |
| **16** | | | | | | | | | | | | | | | | | | | | | |
| **17** | | | | | | | | | | | | | | | | | | | | | |
| **18** | | | | | | | | | | | | | | | | | | | | | |
| **19** | | | | | | | | | | | | | | | | | | | | | |
| **20** | | | | | | | | | | | | | | | | | | | | | |
| **21** | | | | | | | | | | | | | | | | | | | | | |
| **22** | | | | | | | | | | | | | | | | | | | | | |
| **22** | | | | | | | | | | | | | | | | | | | | | |
| **23** | | | | | | | | | | | | | | | | | | | | | |
| **24** | | | | | | | | | | | | | | | | | | | | | |
| **25** | | | | | | | | | | | | | | | | | | | | | |
| tot | | | | | | | | | | | | | | | | | | | | | |
| | **a** | **b** | | **c** | **d** | | **e** | **f** | | **g** | **h** | | **j** | **k** | | **l** | **m** | | **n** | **o** | |

$(NC) = (a + c + e + g + j + l + n) / 7$      $(PC) = (b + d + f + h + k + m + o) / 7$

**FIGURE 5.3**   NC–PC Sheet.

**MAPO WARD SUMMARY**           Date _____

| Hospital: | Ward: | Ward code: |
|---|---|---|

Nr of Beds _____               Nr of Operators (**Op**)   |_____|

Nr of non-cooperative patients **NC** _____      Nr of partially cooperative patients **PC** _____

| LIFTING DEVICES FACTOR (LF) | VALUE OF LF | |
|---|---|---|
| LIFTING AIDS ABSENT OR PRESENT BUT NEVER USED | 4 | |
| ABSENT OR INADEQUATE (% ATL ≤ 90%) + INSUFFICIENT | 4 | |
| Lifting Devices INSUFFICIENT OR INADEQUATE Lifting Devices | 2 | \|_____\| LF |
| ADEQUATE AND SUFFICIENT Lifting Devices | 0.5 | |

| MINOR AIDS FACTOR (AF) | VALUE OF AF | |
|---|---|---|
| Minor Aids ABSENT OR INSUFFICIENT | 1 | |
| Minor Aids SUFFICIENT AND ADEQUATE (% APL ≥ 90%) | 0.5 | \|_____\| AF |

**WHEELCHAIR FACTOR (WF)**

| Mean wheelchair score (MSWh) | 0–1.33 | | 1.34–2.66 | | 2.67-4 | | |
|---|---|---|---|---|---|---|---|
| Numerical sufficiency | NO | YES | NO | YES | NO | YES | \|_____\| WF |
| VALUE OF WF | 1 | 0.75 | 1.5 | 1.12 | 2 | 1.5 | |

**ENVIRONMENT FACTOR**

| Mean environment score (MSENV) | 0 – 5.8 | 5.9 – 11.6 | 11.7 – 17.5 | |
|---|---|---|---|---|
| VALUE OF EF | 0.75 | 1.25 | 1.5 | \|_____\| EF |

| TRAINING FACTOR | VALUE OF TF | |
|---|---|---|
| Adequate training | 0.75 | |
| Only information | 1 | |
| No training | 2 | \|_____\| TF |

**MAPO INDEX**

$$\text{MAPO Index} = (|_____| \times |\_\_\_\_| + |_____| \times |\_\_\_\_|) \times |\_\_\_\_| \times |\_\_\_\_\_| \times |\_\_\_\_| =$$

        NC/OP    LF    PC/OP    AF    WF    EF    TF

| MAPO INDEX | EXPOSURE LEVEL |
|---|---|
| 0 | ABSENT |
| 0.1 – 1.5 | NEGLIGIBLE |
| 1.51 – 5 | MEDIUM |
| >5 | HIGH |

**FIGURE 5.4**    MAPO ward summary.

FONDAZIONE IRCCS CA' GRANDA
OSPEDALE MAGGIORE POLICLINICO
CLINICA DEL LAVORO – MILAN (ITALY)
ERGONOMICS SECTION

## DATA COLLECTION SHEET – SCREENING FOR MANUAL PATIENT HANDLING IN WARDS

| DESCRIPTION OF THE HEALTHCARE FACILITY | | |
|---|---|---|
| HOSPITAL/ RESIDENTIAL CARE FACILITY: | WARD: | WARD CODE: |
| Nr BEDS: | AVERAGE HOSPITAL STAY (days): | DATE: |
| **Nr OF OPERATORS ENGAGED IN MPH:** indicate the total number of operators per job category | | |
| Nursing staff: | Nurses aides: | Other: |
| **MANDATORY ANSWER TO "KEY ENTERS" (QUESTION 1, 2, 3)** | | |

1.　Are there, <u>at least once a day</u> (per operator), push/pull activities with stretchers, beds, wheeled equipment uncomfortable for the operators?
　☐ YES　☐ NO (if YES assess with **SNOOK-CIRIELLO** methodology )

2.　Are there, <u>at least once a day</u> (per operator), manual lifting operations of loads/objects ≥ 10 kg?
　☐ YES　　☐ NO (if YES assess with **NIOSH** methodology)

3.　Are disabled patients usually hospitalized in the ward (NC or PC*) ?
　☐ YES　　☐ NO　　　　　　　　　　**if YES fill in the following fields**

| Nr of OPERATORS ENGAGED IN MPH OVER 3 SHIFTS: indicate the number of operators on duty per shift | | | |
|---|---|---|---|
| SHIFT | **morning** | **Afternoon** | **night** |
| Shift schedule: (00:00 to 00:00) | from_____to_____ | from_____to_____ | from_____to_____ |
| Nr of operators over entire shift | | | |
| **(A)** Total operators over entire shift = | | | |

**Nr of OF PART-TIME OPERATORS:** indicate the exact number of hours worked and calculate them as unit fractions (in relation to the overall duration of the shift).

| Nr of part-time operators present | Hours worked in shift: (00:00 to 00:00) | Unit fraction | (unit fraction by Nr of operators) |
|---|---|---|---|
| | from_____to_____ | | |
| | from_____to_____ | | |
| | from_____to_____ | | |
| **(B)** Total operators (as unit fractions) present by shift duration = | | | |
| **TOTAL Nr OF OPERATORS ENGAGED IN MPH OVER 24 HOURS (Op):** add the total number of operators present over the entire shift **(A)** to the total number of part-time operators **(B)** | | | **OP** |

**Is the work carried out by two nurses? If it is, indicate the number of 2-nurse teams per shift:**
1° morning _____ 2° afternoon_____ 3° night _____

| OPERATOR EDUCATION AND TRAINING | | | |
|---|---|---|---|
| **EDUCATION AND TRAINING** | | **INFORMATION** | |
| Attended theoretical/practical course | ☐ YES　☐ NO | Training only on how to use equipment | ☐ YES　☐ NO |
| if YES, how many months ago? and how many hours/operator | Months _____ hours _____ | Only provided brochures on MPH | ☐ YES　☐ NO |
| if YES, how many operators? | | if YES, how many operators? | |
| Was EFFECTIVENESS measured and documented in writing? | ☐ YES | ☐ NO | |

**FIGURE 5.5**　Screening MAPO. *(continued)*

| TYPE OF PATIENTS: |
|---|

* "Totally Non-Cooperative" patients (**NC**) are patients who <u>need to be fully lifted</u> in transfer/repositioning operations.
"Partially Cooperative" patients (**PC**) are patients who need <u>only partial lifting</u>.

DISABLED PATIENTS (**D**)_____ (indicate average number per day)
Non-Cooperative patients (**NC**) Nr_____     Partially Cooperative patients (**PC**) Nr _____

| DISABLED PATIENTS | Nr NC | Nr PC |
|---|---|---|
| Elderly with multiple concomitant diseases | | |
| Hemiplegic | | |
| Surgical | | |
| Severe stroke | | |
| Dementia | | |
| Other neurologic diseases | | |
| Fracture | | |
| Bariatric | | |
| Other | | |
| **Total** | | |

| PATIENT HANDLING TASKS CURRENTLY CARRIED OUT IN ONE SHIFT: | | | | | | |
|---|---|---|---|---|---|---|
| **MANUAL HANDLING:** describe routine tasks involving total or partial patient lifting | Total lifting (TL) WITHOUT EQUIPMENT | | | Partial Lifting (PL) WITHOUT EQUIPMENT | | |
| indicate the **number of tasks** per shift involving manual patient handling | morning | afternoon | night | morning | afternoon | night |
| | A | B | C | D | E | F |
| ☐ **pulling up in bed** | ☐☐☐☐ | ☐☐☐☐ | ☐☐☐☐ | ☐☐☐☐ | ☐☐☐☐ | ☐☐☐☐ |
| ☐ **turning over in bed (to change position)** | | | | ☐☐☐☐☐ | ☐☐☐☐☐ | ☐☐☐☐☐ |
| ☐ **bed-to-wheelchair and viceversa** | ☐☐ | ☐☐ | ☐☐ | ☐☐ | ☐☐ | ☐☐ |
| ☐ **lifting from seated to upright position** | | | | ☐☐ | ☐☐ | ☐☐ |
| ☐ **bed-to-stretcher and viceversa** | ☐☐ | ☐☐ | ☐☐ | ☐☐ | ☐☐ | ☐☐ |
| ☐ **wheelchair-to-toilet and viceversa** | ☐☐ | ☐☐ | ☐☐ | ☐☐ | ☐☐ | ☐☐ |
| ☐ **other** | ☐☐ | ☐☐ | ☐☐ | ☐☐ | ☐☐ | ☐☐ |
| ☐ **other** | ☐☐ | ☐☐ | ☐☐ | ☐☐ | ☐☐ | ☐☐ |
| TOTAL: calculate the total for each column | | | | | | |
| **Number of total (TL) or partial (PL) manual lifting tasks** | A+B+C = TL | | | D+E+F=PL | | |

| AIDED HANDLING: describe routine tasks involving total or partial patients lifting using available equipment | Total lifting (TL) AIDED | | | Partial Lifting (PL) AIDED | | |
|---|---|---|---|---|---|---|
| Indicate the **number of tasks** per shift involving aided patient handling | morning | afternoon | night | morning | afternoon | night |
| | G | H | I | L | M | N |
| ☐ **pulling up in bed** | ☐☐☐☐ | ☐☐☐☐ | ☐☐☐☐ | ☐☐☐☐ | ☐☐☐☐ | ☐☐☐☐ |
| ☐ **turning over in bed (to change position)** | | | | ☐☐☐☐☐ | ☐☐☐☐☐ | ☐☐☐☐☐ |
| ☐ **bed-to-wheelchair and viceversa** | ☐☐ | ☐☐ | ☐☐ | ☐☐ | ☐☐ | ☐☐ |
| ☐ **lifting from seated to upright position** | | | | ☐☐ | ☐☐ | ☐☐ |
| ☐ **bed-to-stretcher and viceversa** | ☐☐ | ☐☐ | ☐☐ | ☐☐ | ☐☐ | ☐☐ |
| ☐ **wheelchair-to-toilet and viceversa** | ☐☐ | ☐☐ | ☐☐ | ☐☐ | ☐☐ | ☐☐ |
| ☐ **other** | ☐☐ | ☐☐ | ☐☐ | ☐☐ | ☐☐ | ☐☐ |
| ☐ **other** | ☐☐ | ☐☐ | ☐☐ | ☐☐ | ☐☐ | ☐☐ |
| TOTAL: calculate the total for each column | | | | | | |
| **AIDED handling total (ATL) or partial (APL) lifting** | G+H+I = ATL | | | L+M+N=APL | | |
| **% OF AIDED TOTAL LIFTING OPERATIONS (% ATL)** | ATL (TL + ATL) | | | | | |
| **% OF AIDED PARTIAL LIFTING OPERATIONS (% APL)** | | | | APL (PL + APL) | | |

**FIGURE 5.5** *(continued)*   Screening MAPO. *(continued)*

**EQUIPMENT FOR DISABLED PATIENT LIFTING/TRANSFER \***

| EQUIPMENT DESCRIPTION | | Nr | Lack of essential requirements | | Lack of adaptability to patients or environment | | Lack of maintenance | |
|---|---|---|---|---|---|---|---|---|
| LIFTING EQUIPMENT type: | | | YES | NO | YES | NO | YES | NO |
| LIFTING EQUIPMENT type : | | | YES | NO | YES | NO | YES | NO |
| LIFTING EQUIPMENT type : | | | YES | NO | YES | NO | YES | NO |
| Adjustable STRETCHER type : | | | YES | NO | YES | NO | YES | NO |
| Adjustable STRETCHER type : | | | YES | NO | YES | NO | YES | NO |

**OTHER AIDS (MINOR AIDS):**

| EQUIPMENT DESCRIPTION | | Nr | Lack of essential requirements | | Lack of adaptability to patients or environment | | Lack of maintenance | |
|---|---|---|---|---|---|---|---|---|
| SLIDING SHEETS | | | YES | NO | YES | NO | YES | NO |
| STANDING HOISTS type: | | | YES | NO | YES | NO | YES | NO |
| ERGONOMIC BELTS: | | | YES | NO | YES | NO | YES | NO |
| SLIDING BOARDS: | | | YES | NO | YES | NO | YES | NO |
| OTHER: | | | YES | NO | YES | NO | YES | NO |

\* N.B. : Attach floor plan to assess available space for more equipment and if there is an equipment storage room

**FIGURE 5.5** *(continued)* Screening MAPO. *(continued)*

FONDAZIONE IRCCS CA' GRANDA
OSPEDALE MAGGIORE POLICLINICO
CLINICA DEL LAVORO – MILAN (ITALY)
ERGONOMICS SECTION

### SCREENING MAPO WARD SUMMARY                    Date _____

| Hospital / Residential care facility: | Ward: | Ward code: |
|---|---|---|

Nr of Beds _____                                    Nr of Operators (**Op**)    |____|

Nr of non-cooperative patients **NC** _____        Nr of partially cooperative patients **PC** _____

| LIFTING DEVICES FACTOR (LF) | VALUE OF LF | |
|---|---|---|
| LIFTING AIDS ABSENT OR PRESENT BUT NEVER USED | 4 | |
| ABSENT OR INADEQUATE (% ATL ≤ 90%) + INSUFFICIENT Lifting Devices | 4 | |____| LF |
| INSUFFICIENT OR INADEQUATE Lifting Devices | 2 | |
| ADEQUATE AND SUFFICIENT Lifting Devices | 0.5 | |

| MINOR AIDS FACTOR (AF) | VALUE OF AF | |
|---|---|---|
| Minor Aids ABSENT OR INSUFFICIENT | 1 | |____| AF |
| Minor Aids SUFFICIENT AND ADEQUATE (% APL ≥ 90%) | 0.5 | |

| *WHEELCHAIR FACTOR (WF)* | HOSPITAL |__1,5__| WF | RESIDENTIAL CARE FACILITY |__1__| WF |
|---|---|---|

| *ENVIRONMENT FACTOR (EF)* | HOSPITAL |__1,25__| WF | RESIDENTIAL CARE FACILITY |__1__| EF |
|---|---|---|

| TRAINING FACTOR | VALUE OF TF | |
|---|---|---|
| Adequate training | 0.75 | |
| Only information | 1 | |____| TF |
| No training | 2 | |

### MAPO INDEX SCREENING

MAPO Residential $= \left(|\_\_\_| \times |\_\_\_| + |\_\_\_| \times |\_\_\_|\right) \times |\underline{\textbf{1}}| \times |\underline{\textbf{1}}| \times |\_\_\_| = |\_\_|\_\_|$
                  NC/OP    LF    PC/OP   AF            WF         EF      TF      MAPO

MAPO Hospital $= \left(|\_\_\_| \times |\_\_\_| + |\_\_\_| \times |\_\_\_|\right) \times |\underline{\textbf{1,5}}| \times |\underline{\textbf{1,25}}| \times |\_\_\_| = |\_\_|\_\_|$
               NC/OP    LF    PC/OP   AF              WF          EF           TF      MAPO

| MAPO INDEX | EXPOSURE LEVEL |
|---|---|
| 0 | ABSENT |
| 0,1 – 1,5 | NEGLIGIBLE |
| 1,51 – 5 | MEDIUM |
| >5 | HIGH |

**FIGURE 5.5** *(continued)*    Screening MAPO.

# 6 MAPO Index Validation
## Studies on the Association with WMSDs

## 6.1 INTRODUCTION

In order to effectively interpret the significance of studies looking into the relationship between the MAPO (*movimentazione e assistenza pazienti ospedalizzati*—movement and assistance for hospitalized patients) risk index and acute low-back pain in hospital wards, it is worth remembering that this methodological proposal was originally developed in the period from 1994 to 1997. Since then, healthcare systems have undergone massive changes, as illustrated elsewhere in this volume.

As in all occupational health and safety studies, the approach initially consisted of describing the activities carried out by healthcare workers in order to determine how their work was organized and what criticalities were involved. To do this, diaries covering 24-hour periods were drafted and completed for around 200 different hospital wards and elderly care homes. In brief, data concerning the following activities were collected:

- Total or partial patient lifting and handling
- The organization of patient-handling activities and definition of staffing numbers and shifts
- Activities involving awkward postures
- The handling of loads and objects
- Pushing and pulling activities

By analyzing the relevant variables, authors identified the numbers and types of handling and moving activities undertaken for different types of patients. The presence of patient-moving devices was reported, as well as specific information about the positions that certain patients had to adopt for treatment purposes or due to specific clinical situations.

Working postures deemed to be awkward were examined in relation to a number of different factors, such as furnishings, spaces, type of devices used, and training to cope with specific risks. In this respect, therefore, analyzing the aforementioned aspects was a way of assessing them, albeit indirectly.

The handling of loads and pushing/pulling activities were seldom reported, and the risk index values only rarely suggested the presence of risk.

Following this very lengthy observation period, the aim of the study was similar to that of the investigation by the National Institute for Occupational Safety and Health (NIOSH) into the manual handling of loads. It consisted of developing an exposure risk index encompassing all the information required to identify criticalities requiring corrective action to be dealt with in order of priority.

In hospitals, as in other work environments, it is very helpful to employ simple tools providing clear-cut, straightforward information to steer decisions and to generate measurable and comparable results. However, while possibly somewhat flawed due to the extensive variability encountered in a wide range of different work scenarios, the approach traditionally pursued by the International Organization for Standardization (ISO) does utilize parametric methods for assessing workplace risk.

To achieve this objective, therefore, data had to be converted into parameters, and a mathematical model needed to be defined that could readily represent the level of risk exposure associated with manual patient handling. Once the parameters were developed for various risk factors, the next step was to assess whether the outcomes led to the identification of incremental exposure levels and what relationship these had with acute lumbar injury.

This assessment was made by the Unità di Ricerca Ergonomia della Postura e del Movimento (Ergonomics of Posture and Movement [EPM] Research Unit), which, in 1997, after defining the MAPO risk index (in Italian, *Movimentazione e Assistenza a Pazienti Ospedalieri*—Movement and Assistance of Hospital Patients), embarked on a multicenter study to validate the method; the results of the study were published in *La Medicina del Lavoro* (Menoni et al. 1999).

The MAPO index was used to detect risk levels in 222 wards, where 3,440 exposed nursing staff were given clinical/functional spinal exams (Occhipinti 1989) to check for low-back injury. The initial analysis into the links between exposure levels and acute lumbar injury was encouraging, leading to the preliminary definition of risk classes based on the well-known "traffic light" model: For MAPO index levels ranging from 0 to 1.5, the risk level is deemed to be negligible (green); for levels of between 1.51 and 5.00, the risk level is average (yellow) and the annual prevalence of acute low-back pain is around 2.4 times that measured in the risk absent/negligible risk class. For MAPO index values above 5.00 (red), the risk level is regarded as high and the annual prevalence of low-back pain was 5.6 times higher than in the absent/negligible risk class.

However, to confirm the validity of the risk assessment method, the research needed to be expanded; therefore, in 2000 and 2001, the EPM research unit coordinated another multicenter trial involving 23 Italian hospitals, 203 wards, and 3,063 exposed workers. This latest study, published in 2012 by Battevi, Menoni, and Alvarez-Casado, also tested the association between the MAPO screening method and acute low-back pain. The results will be disclosed later in this chapter.

The participants in each of the multicenter studies were given preliminary training to illustrate both the methods for collecting the MAPO risk exposure index data and for identifying low-back pain. During these sessions, investigators were also given electronic devices for controlled data collection purposes. The preliminary meeting was followed by many more, which took place on-site at the various hospitals participating in the study, during which the degree of compliance with the proposed protocols was evaluated. As a result of these evaluations, some data were discarded. Data relating to two variables (exposure and injury) were collected almost concurrently in two periods: 1997–1999 and 2000–2001.

The SPSS statistical software package was used for the description analysis, while the link between exposure and injury was assessed using the STATA 6.0 statistical package. The response variable (acute lumbar injury) was considered as binary for each subject participating in the study: presence of injury (one or more episodes of acute low-back pain) or absence of injury (no episodes).

The odds ratios for three incremental exposure levels were then calculated (MAPO index between 1.51 and 5, between 5.01 and 10, and over 10) using as a reference the MAPO index level corresponding to a value between 0 and 1.5 and therefore with no or negligible exposure in relation to manual patient handling. To begin, a crude analysis was carried out, followed by multivariate analyses to assess the effects of potential confounding factors (i.e., gender, age, and job seniority).

The assumption underlying the decision to opt for this analysis method included the hypothesis that psychosocial factors might be evenly spread over exposed and nonexposed subjects, and therefore the study variables were influenced only by aspects relating to biomechanical overload of the spine.

## 6.2 RESULTS AND DESCRIPTION OF EXPOSURE LEVELS IN STUDY WARDS

Table 6.1 indicates the most commonly represented wards in the two studies, broken down by MAPO index exposure class. It can be noted that in the ward samples,

## TABLE 6.1
## Types of Wards and Distribution per MAPO Risk Class

| | MAPO Exposure Classes | | | | | | | | |
| --- | --- | --- | --- | --- | --- | --- | --- | --- | --- |
| | 0–1.5 | | 1.51–5 | | 5.01–10 | | >10 | | Total |
| Type of Ward | No. | % | No. | % | No. | % | No. | % | No. |
| Medicine 1999 | 2 | 6.7 | 6 | 20 | 13 | 43 | 9 | 30 | 30 |
| Medicine 2003 | 9 | 16 | 23 | 42 | 14 | 25 | 9 | 16 | 55 |
| Surgery 1999 | 3 | 18 | 6 | 35 | 3 | 18 | 5 | 29 | 17 |
| Surgery 2003 | 7 | 19 | 15 | 41 | 8 | 22 | 7 | 19 | 37 |
| OB-GYN 1999 | 1 | 11 | 5 | 56 | 3 | 33 | — | — | 9 |
| OB-GYN 2003 | 2 | 50 | 2 | 50 | — | — | — | — | 4 |
| Orthopedics 1999 | 2 | 14 | 2 | 14 | 4 | 29 | 6 | 43 | 14 |
| Orthopedics 2003 | 3 | 18 | 4 | 24 | 4 | 24 | 6 | 35 | 17 |
| Geriatrics 1999 | 3 | 2.8 | 31 | 29 | 31 | 29 | 44 | 41 | 109 |
| Geriatrics 2003 | — | — | 5 | 42 | 5 | 42 | 2 | 17 | 12 |
| Neurology 1999 | — | — | — | — | 1 | | 1 | | 2 |
| Neurology 2003 | — | — | 6 | 86 | 1 | 14 | — | — | 7 |
| Total | 32 | 10 | 105 | 33 | 86 | 27 | 89 | 28 | 313 |

**TABLE 6.2**

**Analysis of Individual Risk Factors Associated with Manual Patient Lifting**

| Factor | Sufficient and Adequate | | Inadequate or Insufficient | | Absent or Completely Inadequate | |
|---|---|---|---|---|---|---|
| | No. | % | No. | % | No. | % |
| Lifting factor (LF) | 28 | 15 | 44 | 23 | 120 | 62 |
| Minor aids factor(AF) | 3 | 1.5 | — | — | 192 | 98.5 |
| Environment factor (EnvF) | 48 | 23.6 | 133 | 65.5 | 22 | 10.8 |
| Wheelchair factor (WF) | 133 | 65.5 | 65 | 32 | 5 | 2.5 |
| Training factor (TF) | 14 | 6.9 | 33 | 16.3 | 156 | 76.8 |

negligible risk levels are seldom present in either study, while the highest risk level can be found in wards where patients are most often likely to be completely non-cooperative (NC) (e.g., geriatrics and orthopedics).

Overall, 90% of the wards featured non-negligible risk and, of these, many were in a high-risk category. Unexpectedly, risk was detected in certain pediatric wards, since patients—at least in Italy—may be up to 18 years of age.

The individual risk factors listed in Table 6.2 for the 2003 multicenter study are of particular interest insofar as they suggest a dire lack of preventive strategies for reducing risk levels. The 1999 multicenter trial indicated that the use of aids such as patient-lifting devices was not widespread and that, even when provided (in 22.6% of wards), such devices were generally inadequate for meeting the needs of the ward. Even training, which plays such a crucial role in preventive strategies, proved to be inadequate, when provided at all.

An analysis of the individual risk factors associated with the MAPO index level in the various wards may form the basis for a risk reduction plan enabling intervention priorities to be set along with specific actions for reducing exposure levels. Table 6.2 shows that there were virtually no minor aids in the ward sample (e.g., slide sheets, transfer boards, etc.); moreover, training to reduce specific risks was extremely poor.

In 62.5% of wards housing noncooperative patients depending entirely on care-givers to move them, there were either no lifting devices at all or those present failed to meet the ward's needs. Fortunately, the situation improved somewhat when the wheelchair and environmental factors were analyzed. The environmental factor is of particular importance, since every structural intervention involves a major financial investment.

Tables 6.3, 6.4, 6.5, and 6.6 describe the main characteristics of the exposed subjects included in the study. There is a distinct prevalence of female workers (the male-to-female ratio is 1:4), younger subjects, and individuals with relatively low job seniority. Table 6.4 indicates that most of the workers were aged between 26 and 55 (91.9%). The above-55 age class was poorly represented.

A comparison between Tables 6.5 and 6.6 shows that there are major differences within the sample between the different classes of ward and job seniority. Notably, 56.4% of the subjects had worked in the ward for less than 4 years; conversely, there

**TABLE 6.3**

**Mean and Standard Deviation for Sample Population Broken Down by Age, Job, Ward Seniority, and Gender**

|  | x | SD |
|---|---|---|
| **Males (645)** | | |
| Age | 36.8 | 8.7 |
| Number of years in ward | 6.1 | 7.1 |
| Number of years in job | 11.0 | 8.4 |
| **Females (2,418)** | | |
| Age | 36.0 | 8.4 |
| Number of years in ward | 5.8 | 5.9 |
| Number of years in job | 10.4 | 7.1 |

**TABLE 6.4**

**Breakdown of Sample Population by Age and Gender**

| | Age Class (Years) | | | | | | | | | | Total |
|---|---|---|---|---|---|---|---|---|---|---|---|
| | <25 | | 26–35 | | 36–45 | | 46–55 | | >55 | | |
| Gender | No. | % | No. | % | No. | % | No. | % | No. | % | No. |
| Male | 26 | 4.0 | 317 | 49.1 | 175 | 27.1 | 113 | 17.6 | 14 | 2.2 | 645 |
| Female | 151 | 6.2 | 1172 | 48.5 | 745 | 30.8 | 293 | 12.1 | 57 | 2.4 | 2418 |
| Total | 177 | 5.8 | 1489 | 48.6 | 920 | 30.1 | 406 | 13.2 | 71 | 2.3 | 3063 |

**TABLE 6.5**

**Ward Seniority by Gender**

| | Gender | | | | | |
|---|---|---|---|---|---|---|
| Ward Seniority Class | Male | | Female | | Total | |
| (No. Years in Ward) | No. | % | No. | % | No. | % |
| 0–4 | 373 | 57.8 | 1354 | 56.0 | 1727 | 56.4 |
| 5–9 | 129 | 20.0 | 526 | 21.8 | 655 | 21.4 |
| 10–14 | 61 | 9.5 | 312 | 12.9 | 373 | 12.2 |
| >14 | 82 | 12.7 | 226 | 9.3 | 308 | 10.0 |
| Total | 645 | 100.0 | 2418 | 100.0 | 3063 | 100.0 |

**TABLE 6.6**
**Job Seniority by Gender**

| Job Seniority Class (No. Years in Job) | Gender | | | | | |
| | Male | | Female | | Total | |
| | No. | % | No. | % | No. | % |
|---|---|---|---|---|---|---|
| 0–4 | 161 | 25.0 | 505 | 20.9 | 666 | 21.7 |
| 5–9 | 207 | 32.1 | 757 | 31.3 | 964 | 31.5 |
| 10–14 | 93 | 14.4 | 549 | 22.7 | 642 | 21.0 |
| >14 | 184 | 28.5 | 605 | 25.1 | 789 | 25.8 |
| Total | 645 | 100.0 | 2416[a] | 100.0 | 3061[a] | 100.0 |

[a] Two missing.

were no major differences between the various classes with regard to job seniority. This suggests a fairly high turnover rate.

Table 6.7 describes the distribution of the workers by type of ward; as mentioned previously, the most common specialties are medicine, surgery, orthopedics, and geriatrics.

Table 6.8 indicates that the workers in the sample who performed patient-handling activities had various qualifications.

**TABLE 6.7**
**Distribution of Exposed Population by Ward Type**

| Ward | Gender | | |
| | No. Males | No. Females | Total |
|---|---|---|---|
| Medicine | 232 | 865 | 1097 |
| Surgery | 136 | 420 | 556 |
| Cardiology | 25 | 88 | 113 |
| Coronary intensive care | 14 | 25 | 39 |
| Infectious diseases | 16 | 28 | 44 |
| Neurology | 26 | 98 | 124 |
| OB-GYN | — | 46 | 46 |
| Orthopedics | 69 | 204 | 273 |
| Pneumology | 10 | 25 | 35 |
| ENT | 8 | 23 | 31 |
| Pediatrics–neonatology | 1 | 45 | 46 |
| Urology | 20 | 11 | 31 |
| Nephrodialysis | — | 29 | 29 |
| Geriatrics | 10 | 224 | 234 |
| Other | 78 | 287 | 365 |
| Total | 645 | 2418 | 3063 |

---

**TABLE 6.8**
**Breakdown of Sample Population**
**by Specific Job**

| Job | No. | % |
|---|---|---|
| Registered nurse | 1993 | 65 |
| Orderly/attendant | 535 | 17.5 |
| Nurse's aide | 279 | 9.1 |
| General nurse | 143 | 4.7 |
| Head nurse | 82 | 2.7 |
| Pediatric nurse | 22 | 0.7 |
| Midwife | 5 | 0.2 |
| Missing | 4 | 0.1 |
| Total | 3063 | 100 |

---

In recent years, there has been an increase in the percentage of orderlies and attendants to offset the shortage of specialized nursing staff.

## 6.3 RELATIONSHIP BETWEEN MAPO EXPOSURE INDEX AND ACUTE LOW-BACK PAIN IN THE PREVIOUS 12 MONTHS

Based on the exclusion criteria employed, data were processed concerning 191 wards and 2,602 subjects identified as being exposed to patient-handling activities. Table 6.9 shows the association between acute lumbar pain in the previous year and the MAPO index. The results show that the crude odds ratios for exposure levels (MAPO index) above 1.5 were all positive and statistically significant with respect to MAPO ≤ 1.5. Moreover, for the second and third exposure classes, the trend was upward and tended to decline or stabilize at the highest exposure level (MAPO > 10). The odds ratios, adjusted for the confounding factors of gender, age, and total job seniority, were not significantly different. These findings are virtually identical to those that emerged in the previous multicenter study (Battevi et al. 1999).

The data raised an interesting question: Despite the association between acute low-back pain and increasing MAPO index levels, why was a decrease observed in class 4 with an exposure index of >10?

The question was at least partly answered when the results of the previous multicenter study were presented. According to the most likely hypothesis, in situations of significant exposure to manual patient lifting, the behavior of the workers and the exposure level did not correspond (i.e., the exposure was misclassified). In other words, when patient-handling tasks were excessively demanding, the operator was unable to perform them completely and therefore the effort might not correspond to the MAPO index level. Should this hypothesis be accurate, then we might observe different behaviors between medical wards and surgical wards. The need for urgent patient handling in critical surgical situations could determine a lower exposure misclassification risk and therefore a better match between the MAPO index (that

**TABLE 6.9**

**Association between the MAPO Index and the Occurrence of Acute Low-Back Pain in the Previous Year**

| | Acute Low-Back Pain in the Previous Year | | Odds Ratio | CI (95%) | Odds Ratio (Adjusted) | CI (95%) (Adjusted) |
|---|---|---|---|---|---|---|
| | Neg. | Pos. | | | | |
| **MAPO Index** | | | | | | |
| 0–1.5 | 338 | 19 | 1 | — | 1 | — |
| 1.51–5 | 1024 | 140 | 2.43 | 1.43–3.99 | 2.36 | 1.43–3.87 |
| 5.01–10 | 515 | 93 | 3.21 | 1.91–5.39 | 3.13 | 1.87–5.24 |
| >10 | 407 | 67 | 2.92 | 1.71–5.00 | 2.83 | 1.66–4.82 |
| **Gender** | | | | | | |
| Male | 470 | 67 | 1 | — | — | — |
| Female | 1814 | 252 | 0.97 | 0.73–1.22 | 1.02 | 0.76–1.37 |
| **Age Class (Years)** | | | | | | |
| 15–25 | 112 | 15 | 1 | — | | |
| 26–35 | 1105 | 126 | 0.85 | 0.48–1.50 | 0.82 | 0.45–1.49 |
| 36–45 | 696 | 114 | 1.32 | 0.68–2.17 | 1.25 | 0.67–2.33 |
| 46–55 | 313 | 158 | 1.38 | 0.75–2.54 | 1.47 | 0.75–2.90 |
| >55 | 58 | 6 | 0.77 | 0.28–2.10 | 0.86 | 0.30–2.46 |
| **Job Seniority (No. Years in Job)** | | | | | | |
| 0–4 | 422 | 54 | 1 | — | — | — |
| 5–9 | 737 | 98 | 0.92 | 0.60–1.43 | 0.97 | 0.67–1.40 |
| 10–14 | 500 | 75 | 1.05 | 0.67–1.64 | 1.01 | 0.68–1.51 |
| >14 | 624 | 92 | 1.00 | 0.65–1.54 | 0.82 | 0.53–1.26 |

estimates the extent of the future handling task) and the handling task actually performed in the specific ward.

To test the hypothesis, an analysis was conducted of the association between the MAPO index and the presence of acute low-back pain in the previous 12 months, based on two ward subsets: medical wards and surgical wards. The analysis involved 108 medical wards with a total of 1,500 exposed subjects, and 76 surgical wards with a total of 840 exposed subjects. Tables 6.10 and 6.11 illustrate the analysis.

Comparing these two tables, it can be seen that there is a difference in the trend for the level of association between the MAPO index and acute low-back pain in the previous year: Whereas in medical wards (Table 6.10) there is a strong association for all index classes, with a marked increase in classes 2 and 3 and a decline in class 4, in surgical wards (Table 6.11) the odds ratio increases as the exposure index rises.

As for the other findings, the confounding factors had no major influence on the overall situation. Therefore, the results appear to confirm the validity of the assumption.

**TABLE 6.10**

**Association between the MAPO Index and the Occurrence of Acute Low-Back Pain in the Previous Year—Study Results for Medical Wards**

| | Acute Low-Back Pain in the Previous Year | | Odds Ratio | CI (95%) | Odds Ratio (Adjusted) | CI (95%) (Adjusted) |
|---|---|---|---|---|---|---|
| | Neg. | Pos. | | | | |
| **MAPO Index** | | | | | | |
| 0–1.5 | 120 | 5 | 1 | — | 1 | — |
| 1.51–5 | 680 | 98 | 3.45 | 1.37–8.71 | 3.45 | 1.37–8.69 |
| 5.01–10 | 266 | 55 | 4.96 | 1.90–12.90 | 4.92 | 1.91—12.67 |
| >10 | 247 | 29 | 2.81 | 1.05–7.51 | 2.87 | 1.07–7.65 |
| **Gender** | | | | | | |
| Male | 242 | 31 | 1 | — | — | — |
| Female | 1071 | 156 | 1.13 | 0.75–1.71 | 1.14 | 0.75–1.73 |
| **Age Class (Years)** | | | | | | |
| 15–25 | 72 | 11 | 1 | — | | |
| 26–35 | 646 | 73 | 0.73 | 0.37–1.45 | 0.75 | 0.37–1.53 |
| 36–45 | 400 | 64 | 1.04 | 0.52–2.08 | 1.06 | 0.50–2.25 |
| 46–55 | 161 | 36 | 1.46 | 0.70–3.04 | 1.48 | 0.66–3.32 |
| >55 | 34 | 3 | 0.57 | 0.14–2.22 | 0.61 | 0.15–2.47 |
| **Job Seniority (No. Years in Job)** | | | | | | |
| 0–4 | 282 | 33 | 1 | — | — | — |
| 5–9 | 434 | 60 | 1.21 | 0.68–2.15 | 1.12 | 0.70–1.79 |
| 10–14 | 292 | 45 | 1.27 | 0.70–2.30 | 1.17 | 0.70–1.96 |
| >14 | 304 | 49 | 1.31 | 0.73–2.34 | 1.08 | 0.62–1.87 |

## 6.4 MAPO SCREENING METHOD: VALIDATION STUDY

This research was conducted in the period from 2008 to 2009 (Battevi et al. 2012), with the aim of designing a much-needed rapid risk mapping tool that could be used to plan subsequent research based on certain priorities. Rather than using an analytical methodology (cf. the previous chapter) involving the calculation of the MAPO index via two separate steps, an interview and an on-site inspection, this screening tool requires only an interview to estimate risk in the ward. From the practical standpoint, the wheelchair and environment factors were set at one, eliminating their influence on the calculation of the MAPO index (Table 6.12).

The study was conducted in 31 wards at 10 different residential care facilities (*Residenze Socio Assistenziali* [RSA]), located in Italy's northeastern Veneto region. The same methodology as in the previous multicenter studies was adopted. Data were collected for exposed subjects (411) and for a nonexposed reference sample

**TABLE 6.11**

**Association between the MAPO Index and Acute Low-Back Pain in the Previous Year—Results for Surgical Wards**

| | Acute Low-Back Pain in the Previous Year | | Odds Ratio | CI (95%) | Odds Ratio (Adjusted) | CI (95%) (Adjusted) |
|---|---|---|---|---|---|---|
| | Neg. | Pos. | | | | |
| **MAPO Index** | | | | | | |
| 0–1.5 | 162 | 13 | 1 | — | 1 | — |
| 1.51–5 | 212 | 23 | 1.43 | 0.70–2.90 | 1.33 | 0.65–2.73 |
| 5.01–10 | 208 | 33 | 2.82 | 1.44–5.51 | 1.94 | 0.98–3.85 |
| >10 | 152 | 37 | 3.27 | 1.63–6.56 | 2.87 | 1.46–5.65 |
| **Gender** | | | | | | |
| Male | 185 | 30 | 1 | — | — | — |
| Female | 549 | 76 | 0.85 | 0.54–1.34 | 0.92 | 0.57–1.47 |
| **Age Class (Years)** | | | | | | |
| 15–25 | 30 | 1 | 1 | — | | |
| 26–35 | 342 | 44 | 3.85 | 0.50–29.21 | 0.82 | 0.47–28.13 |
| 36–45 | 226 | 39 | 5.17 | 0.67–39.39 | 1.25 | 0.72–45.42 |
| 46–55 | 118 | 19 | 4.83 | 0.60–38.41 | 1.47 | 0.72–52.27 |
| >55 | 18 | 3 | 5.00 | 0.44–55.70 | 0.86 | 0.69–89.46 |
| **Job Seniority (No. Years in Job)** | | | | | | |
| 0–4 | 94 | 15 | 1 | — | — | — |
| 5–9 | 251 | 32 | 0.55 | 0.24–1.24 | 0.69 | 0.35–1.35 |
| 10–14 | 147 | 26 | 0.95 | 0.42–2.14 | 0.82 | 0.39–1.69 |
| >14 | 242 | 33 | 0.66 | 0.30–1.44 | 0.51 | 0.23–1.13 |

**TABLE 6.12**

**Characteristics of the Risk Factors Examined Using the MAPO Screening Method**

| Factor | Sufficient and Adequate | | Inadequate or Insufficient | | Absent or Completely Inadequate | |
|---|---|---|---|---|---|---|
| | No. | % | No. | % | No. | % |
| Lifting factor (LF) | 17 | 54 | 13 | 41.9 | 1 | 3.2 |
| Minor aids factor (AF) | 3 | 9.7 | — | — | 28 | 90.3 |
| Training factor (TF) | 10 | 32.3 | 13 | 41.9 | 8 | 25.8 |

(237) in the period from 2008 to 2009. The reference population consisted of clerical workers who used a computer for at least 20 hours per week in different offices (town halls, hospitals, and law courts) and were not exposed to load handling and lived in the districts where the facilities were located. The study excluded subjects who had worked in the wards for less than 6 months or in wards in which no acute low-back pain data had been collected for at least 70% of the exposed workers.

The SPSS statistical package was used for data description analysis of both the entire sample population and the external reference sample. The association between exposure and injury (acute low-back pain) was studied using the unconditional logistic model and STATA 6.0 statistical analysis software. For each subject included in the study, the response variable (acute low-back pain) was considered as binary: presence of injury (one or more episodes of acute low-back pain) or absence of injury (no episodes). This analysis was limited to exposed subjects working through three shifts over 24 hours, for at least 30 hours per week (178 subjects).

The odds ratios were then calculated, both crude and adjusted for gender and age class, for increasing exposure levels with respect to the external reference sample.

### 6.4.1 RESULTS AND DESCRIPTION OF EXPOSURE LEVELS IN STUDY WARDS

The exposed subjects in all of the hospitals in the study were all qualified healthcare workers who had completed 600 hours of education and training. There were no professional registered nurses. The 31 wards each had between 15 and 51 beds, and the patients were both NC and PC.

After screening, the average MAPO index level was 4.3 with a standard deviation of 2.9 and a range of between 1.2 and 13.8. Only two wards (6.5%) displayed negligible risk (MAPO index between 0.1 and 1.50), while 70.9% evidenced an average risk index of between 1.51 and 5; the remaining wards (22.6%) featured high manual patient-handling risk (MAPO index above 5).

Excluding the environment and wheelchair factors, which were not screened for, the other risk factors showed the following distribution:

- Overall, the health status of 648 subjects was assessed, of whom 411 were exposed and 237 not exposed; 17.4% of the subjects were male and 82.6% were female. The average age of the workers was somewhat high (45.36 years), while the distribution by age class was as shown in Table 6.13.
- An interesting difference was detected between the exposed versus non-exposed workers who reported at least one episode of acute low-back pain in the previous years, in terms of the prevalence of lumbar disk herniation, as shown in Table 6.14.

### 6.4.2 RELATIONSHIP BETWEEN THE MAPO SCREENING INDEX AND ACUTE LOW-BACK PAIN DURING THE PREVIOUS 12 MONTHS

The first descriptive analysis showed that not all the subjects working shifts were working 36 hours per week per their employment contract; working hours could

**TABLE 6.13**
**Distribution of Sample Population by Age Class**

| Age Class (in Years) | Male No. | Male % | Female No. | Female % | Total No. | Total % |
|---|---|---|---|---|---|---|
| ≤25 | 11 | 9.7 | 32 | 6.0 | 43 | 6.6 |
| 26–35 | 26 | 23.0 | 93 | 17.4 | 119 | 18.4 |
| 36–45 | 28 | 24.8 | 183 | 34.2 | 211 | 32.6 |
| 46–55 | 36 | 31.9 | 193 | 36.1 | 229 | 35.3 |
| >55 | 12 | 10.6 | 34 | 6.4 | 46 | 7.1 |

**TABLE 6.14**
**Prevalence of Exposed and Nonexposed Subjects Reporting at Least One Episode of Acute Low-Back Pain in the Previous Year with Lumbar Disk Herniation**

|  | Exposed No. | Exposed % | Not Exposed No. | Not Exposed % |
|---|---|---|---|---|
| Subjects reporting at least **one episode of acute low-back pain** in the previous year | 37 | 9.0 | 11 | 4.6 |
| Prevalence of subjects with **herniated disk** | 28 | 6.8 | 7 | 3.0 |

**TABLE 6.15**
**Analysis of the Prevalence of Acute Low-Back Pain by Exposure Level and Number of Hours Worked**

|  | Acute Low-Back Pain in Wards With MAPO Index Between 1.51 and 5 No. | Acute Low-Back Pain in Wards With MAPO Index Between 1.51 and 5 % | Acute Low-Back Pain in Wards With MAPO Index Above 5 No. | Acute Low-Back Pain in Wards With MAPO Index Above 5 % |
|---|---|---|---|---|
| All subjects on three-shift rotation (306) | 17 | 6.9 | 8 | 13.1 |
| Subjects on three-shift rotation working at least 24 hours per week (249) | 15 | 8.2 | 8 | 14.3 |
| Subjects on three-shift rotation working at least 30 hours per week (178) | 12 | 9.7 | 8 | 14.8 |

range anywhere from 18 to 38 hours (Table 6.15). Table 6.16 reports the results of a preliminary crude analysis of the prevalence of acute low-back pain in the previous 12 months, broken down by duration of weekly exposure. This description indicates that the prevalence of acute low-back pain during the previous year increased with the duration of exposure.

**TABLE 6.16**

**Association between the MAPO Screening Indexes and Acute Low-Back Pain in the Previous Year—Results for Subjects Working at least 30 Hours on a Three-Shift Rotation**

| | Acute Low-Back Pain in the Previous Year | | Odds Ratio | CI (95%) | Odds Ratio (Adjusted) | CI (95%) (Adjusted) |
|---|---|---|---|---|---|---|
| | Neg. | Pos. | | | | |
| **Index** | | | | | | |
| 0 | 226 | 11 | 1 | — | 1 | |
| 1.51–5 | 112 | 12 | 2.20 | 0.97–5.14 | 2.22 | — |
| 5.01–10 | 46 | 8 | 3.57 | 1.36–9.37 | 3.77 | 0.88–5.63 |
| P value for trend | | | 0.007 | | 0.010 | 1.33–10.74 |
| **Gender** | | | | | | |
| Male | 16 | 2 | 1 | — | 1 | — |
| Female | 142 | 18 | 2.40 | 0.82–7.03 | 1.76 | 0.57–5.42 |
| **Age Class (Years)** | | | | | | |
| 15–25 | 7 | 1 | 1 | — | 1 | — |
| 26–35 | 30 | 6 | 1.26 | 0.24–6.59 | 0.98 | 0.18–5.33 |
| 36–45 | 56 | 6 | 1.17 | 0.24–5.80 | 0.89 | 0.17–4.64 |
| 46–55 | 62 | 6 | 1.30 | 0.27–6.17 | 1.04 | 0.21–5.15 |
| >55 | 3 | 1 | 2.00 | 0.34–11.70 | 2.58 | 0.42–15.96 |

The data were then analyzed using the unconditional logistic regression model, considering only the exposed subjects working at least 30 hours a week on a three-shift rotation. A positive trend clearly emerges from this analysis for the prevalence of acute low-back pain episodes with respect to the level of exposure; more specifically, the odds ratio for subjects exposed to MAPO index levels of between 1.51 and 5 is almost twice as high as for nonexposed subjects, while in the MAPO index class of over 5, the odds ratio is four times higher and the result remains largely unchanged when possible confounding factors such as gender and age class are entered into the analysis.

These results are consistent with those of the two previous studies, confirming that the MAPO screening method is a valid indicator of patient-handling risk—at least for subjects on a three-shift rotation, working at least 30 hours a week.

## 6.5 REFERENCE VALUES

These studies have generated a number of considerations and arguments: The first is very positive, insofar as while the most recent analysis studied a different sample population than in 1999, the results are almost identical, confirming the validity of the original assumption underlying the definition of the three exposure levels shown in Table 6.17. Accordingly, based on the current level of understanding, the MAPO

**TABLE 6.17**

**Correspondence between MAPO Index and Exposure Level**

| MAPO Index Score | Exposure Level |
|---|---|
| 0 | Absent (white) |
| 0.1–1.5 | Negligible (green) |
| 1.51–5.00 | Present (yellow) |
| >5.00 | High (red) |

index is an accurate indicator of risk, with the ability to predict an effect (i.e., acute low-back pain) among hospital workers. This statement also applies to the MAPO screening method.

Earlier conclusions (Battevi et al. 2006) on the association between the analytical MAPO index and acute low-back pain in the previous 12 months were transposed in this volume and found to be valid also in light of the latest study conducted with the MAPO screening method; the conclusions suggest an evolution that does not, however, disprove the theoretical and practical method adopted in 1999 (Menoni et al.).

It should be stressed that, to date, despite countless different approaches toward the organization of work—for instance, shift-sharing among multiple workers, non-rotation between shifts, and so forth—it is often difficult to define a single standard group of exposed subjects; therefore, a descriptive analysis of the types of handling activities performed may allow the MAPO index to be adapted to different organizational circumstances.

# 7 Analysis of Patient-Handling Tasks in Operating Rooms

## 7.1 INTRODUCTION

The process of detecting exposure to risk when handling patients in operating rooms stems from the same rationale underlying the definition of specific risk exposure levels in patient wards (MAPO: *movimentazione e assistenza pazienti ospedalizzati*—movement and assistance for hospitalized patients); risk exposure levels will largely depend on the types and numbers of procedures that require patient handling. In contrast with the situation in wards, where it is possible to calculate a specific risk index based on an aggregate evaluation of all the risk factors analyzed, in operating rooms the analysis and description of risk factors can only *estimate risk exposure levels.*

As in the case of wards, the descriptive analysis will help to identify criticalities in the operating room instantly and accurately, so as to plan the most effective corrective measures. The risk analysis process entails two steps: an interview and an on-site visit. It is advisable for all the contact people involved to be adequately informed of the purpose of the investigation and the type of information to be collected.

The aim of the interview is to gather information about how the department is organized, and the visit is undertaken to collect more details—primarily concerning equipment and work environments—as well as to confirm the accuracy of the information provided during the interview.

## 7.2 DATA COLLECTION PROTOCOL

In order to acquire all the information needed to describe and assess the relevant risk exposure level, a set of data collection forms has been put together (Figure 7.1)—on the one hand to set manual patient-handling activities within the general workplace context, and on the other to collect useful data and define the main risk factors (Figure 7.2).

Table 7.1 indicates the risk factors identified in the surgical unit; as elsewhere, the main risk factor is represented by the relationship between the number of surgical (NS) procedures requiring patient handling and the number of operators (OPs) performing patient-handling tasks over a 24-hour period. The MAPO methodology also takes into consideration the following risk factors: lifting devices, environment, stretchers, and training.

**TABLE 7.1**

**General Outline of Risk Factor Detection in a Surgical Unit**

| Risk Factors | Description |
|---|---|
| NS/OP | Relationship between *average number of* procedures/day requiring patient handling (NS) and number of operators present over a 24-hour period (OP) |
| Equipment devices factor | Ergonomic inadequacy (% of aided lifts) of disabled (D) patient lifting/moving equipment |
| Training factor | Inadequacy of training on specific risk |
| Environment factor | Ergonomic inadequacy of rooms in which patients are moved/handled (space and furnishings) |
| Stretcher factor | Ergonomic inadequacy of stretchers for moving/transferring D patients |

Using the same rationale as that for wards, data are entered into the form based on an interview with the head of the surgical unit, an analysis of the operators involved in manual patient handling (MPH), the type of routine patient-handling activities performed, and training on the prevention of WMSDs (work-related musculoskeletal disorders) due to specific risks.

The analysis should clearly define how work in the surgical unit is organized, with an indication of the average number of procedures performed each day that require patient handling, under both general and local anesthesia.

The entire process is described to enable the interviewer to calculate the number of maneuvers carried out manually and using devices. The data sheet describes and quantifies patient-handling activities by breaking them down by groups of healthcare operators featuring different professional qualifications.

### 7.2.1 Operators Involved in Patient-Handling Activities

All of the operators involved in patient-handling activities are described (Table 7.2), broken down according to their respective qualifications. This information will be useful for quantifying the actual number of MPH operators over the 24-hour period (= OP). The weekly occupancy is also described for the various operating rooms.

**TABLE 7.2**

**Total Surgical Unit Employees in Data Collection Protocol**

| Total Operators (Staff Performing Manual Patient-Handling Tasks): | | | |
|---|---|---|---|
| Nurses/scrub nurses: | Nurses aides: | Other: | |
| Surgical unit daily working hours: | | Working days: | |
| Operators (Staff Performing Manual Patient-Handling Tasks) Present over 24 Hours: | | | |
| Nurses/scrub nurses: | Nurses aides: | Other: | |
| Indicate the number of operators (OPs) as the sum of operators (in all job categories) involved in patient handling and present over 24 hours | | | **OP** |

## 7.2.2 Description and Number of Surgical Procedures

In surgical units, "disabled patients" are identified based on an analysis of surgical procedures requiring patients to be lifted. The types of manual patient-handling activities routinely carried out by surgical unit staff refer to the task of collecting the patient from the ward (occasionally with the help of ward staff) and transferring the patient from the nonsterile ward (= preoperative holding area) to the sterile operating room (OR).

In some surgical units, primarily orthopedics and neurosurgery, patients also have to be turned from supine to prone and vice versa, determining "peak" biomechanical overloads for this type of lifting. In order to describe routine patient-handling tasks more easily, the average number of surgical procedures per day on patients under general anesthesia (GA) or local anesthesia (LA) are counted only if total or partial patient lifting is necessary. The number of surgeries calling for patient handling is thus represented by the sum of GA + LA.

The number of procedures performed is standard information available in all surgical units and is an accurate and reliable source of basic data to which the types of procedures that may require patient handling can be added as shown in Table 7.3.

In a "virtual" surgical unit with two operating rooms that perform 2,800 procedures/year (i.e., 12 surgeries/day, including 8 under general anesthesia and 4 under local anesthesia) calling for the patient to be moved manually, the specific section of the data collection form will be completed as shown in Table 7.4.

---

**TABLE 7.3**
**Quantification of Surgical Procedures**

| No. of ORs __ No. of procedures/year __ Average no. procedures/day: \|__\|__\| • Average no. of procedures/day under general anesthesia (GA): _____ \|__\|__\| • Average no. of procedures/day under local anesthesia: _____ Average no. of procedures/day not requiring total/partial patient lifting _____ Average no. of procedures/day requiring total/partial patient lifting (LA) _____ | | |
|---|---|---|
| No. of procedures requiring patient handling (GA + LA) | | **NS** |

---

**TABLE 7.4**
**Example of Form with Quantification of Surgical Procedures**

| No. of ORs __2__ No. of procedures/year __2,880__ Average no. procedures/day \|_1_\|_2_\| • Average no. of procedures/day under general anesthesia (GA): _____ \|__\|_8_\| • Average no. of procedures/day under local anesthesia: ___4___ Average no. of procedures/day not requiring total/partial patient lifting _____ Average no. of procedures/day requiring total/partial patient lifting (LA) ___4___ | | |
|---|---|---|
| No. of procedures requiring patient handling (GA + LA) | 12 | **NS** |

**TABLE 7.5**
**Patient-Handling Tasks in a Surgical Unit**

| Type of Anesthesia | GA = No. ____ | | GA = No. ____ | |
|---|---|---|---|---|
| Bed/stretcher | [ ] Manual | [ ] Aided | [ ] Manual | [ ] Aided |
| Stretcher/operating table | [ ] Manual | [ ] Aided | [ ] Manual | [ ] Aided |
| Operating table/stretcher | [ ] Manual | [ ] Aided | [ ] Manual | [ ] Aided |
| Stretcher/bed | [ ] Manual | [ ] Aided | [ ] Manual | [ ] Aided |
| Stretcher/stretcher | [ ] Manual | [ ] Aided | [ ] Manual | [ ] Aided |
| Prone/supine | [ ] Manual | [ ] Aided | [ ] Manual | [ ] Aided |
| Supine/prone | [ ] Manual | [ ] Aided | [ ] Manual | [ ] Aided |
| Column score (no. maneuvers per no. GAs) | ____ A | ____ B | ____ C | ____ D |
| **Type of Anesthesia** | **LA = No. ____** | | **LA = No. ____** | |
| Bed/stretcher | [ ] Manual | [ ] Aided | [ ] Manual | [ ] Aided |
| Stretcher/operating table | [ ] Manual | [ ] Aided | [ ] Manual | [ ] Aided |
| Operating table/stretcher | [ ] Manual | [ ] Aided | [ ] Manual | [ ] Aided |
| Stretcher/bed | [ ] Manual | [ ] Aided | [ ] Manual | [ ] Aided |
| Stretcher/stretcher | [ ] Manual | [ ] Aided | [ ] Manual | [ ] Aided |
| Prone/supine | [ ] Manual | [ ] Aided | [ ] Manual | [ ] Aided |
| Supine/prone | [ ] Manual | [ ] Aided | [ ] Manual | [ ] Aided |
| Column score (no. maneuvers per no. LAs) | ____ E | ____ F | ____ G | ____ H |

### 7.2.3  DESCRIPTION AND QUANTIFICATION OF HANDLING MANEUVERS

Table 7.5 indicates routine patient-handling pathways relating to different types of procedures (GA/LA), based on which it is possible to detect the number of patient-handling activities fairly accurately per method used (i.e., manual or aided).

In general surgery, where procedures are performed on patients coming from a variety of different wards, there may be different handling pathways also depending on what lifting equipment may or may not be on hand in the individual ward. Accordingly, each column in Table 7.5 describes a different manual and aided lifting pathway.

Patient-handling tasks are described to calculate the percentage of tasks performed using equipment (AMPER) that is the ratio of the sum of the column scores for aided maneuvers to the total number of maneuvers performed:

$$\text{Aided maneuvers} = B + D + F + H$$

$$\text{Disabled-patient-handling operations} = A + B + C + D + E + F + G + H$$

#### 7.2.3.1  Example of Description and Quantification of Disabled-Patient-Handling Tasks

In the preceding surgical unit, of the eight procedures performed under general anesthesia, six require moving patients manually from their ward bed to a stretcher and

**TABLE 7.6**

**Example of Patient-Handling Tasks Performed in Surgical Unit**

### Types of Pathway Maneuvers

| | GA = No. 6 | | GA = No. 2 | |
|---|---|---|---|---|
| Bed/stretcher | [x] Manual | [ ] Aided | [ ] Manual | [ ] Aided |
| Stretcher/operating table | [ ] Manual | [x] Aided | [ ] Manual | [x] Aided |
| Operating table/stretcher | [ ] Manual | [x] Aided | [ ] Manual | [x] Aided |
| Stretcher/bed | [x] Manual | [ ] Aided | [ ] Manual | [ ] Aided |
| Stretcher/stretcher | [ ] Manual | [ ] Aided | [ ] Manual | [ ] Aided |
| From prone to supine | [ ] Manual | [ ] Aided | [ ] Manual | [ ] Aided |
| From supine to prone | [ ] Manual | [ ] Aided | [ ] Manual | [ ] Aided |
| Column score (no. maneuvers × no. GAs) | 12 A | 12 B | ___ C | 4 D |

| Type of Anesthesia | LA = No. 4 | | LA = No. ___ | |
|---|---|---|---|---|
| Bed/stretcher | [x] Manual | [ ] Aided | [ ] Manual | [ ] Aided |
| Stretcher/operating table | [ ] Manual | [x] Aided | [ ] Manual | [ ] Aided |
| Operating table/stretcher | [ ] Manual | [x] Aided | [ ] Manual | [ ] Aided |
| Stretcher/bed | [x] Manual | [ ] Aided | [ ] Manual | [ ] Aided |
| Stretcher/stretcher | [ ] Manual | [ ] Aided | [ ] Manual | [ ] Aided |
| From prone to supine | [ ] Manual | [ ] Aided | [ ] Manual | [ ] Aided |
| From supine to prone | [ ] Manual | [ ] Aided | [ ] Manual | [ ] Aided |
| Column score (no. maneuvers × no. LA) | 8 E | 8 F | ___ G | ___ H |

vice versa, while lifting devices are used to move all the patients from stretcher to operating table and vice versa. Thus, two types of "GA pathways" are identified and described in different columns of Table 7.6.

For the four procedures performed under local anesthesia, the patient transfers from ward bed to stretcher and vice versa are carried out manually, while lifting aids are used to move the patient from stretcher to operating table and vice versa.

A situation like this may occur when, for certain procedures, surgical unit staff also transfer patients from the ward to the operating room; for other surgeries, the patients are transferred by ward staff. The overall calculation of the percentage of aided tasks (in this example) will thus be 54%, meaning that only 24 patient-handling tasks (for a total of 44 handling tasks) are aided:

$$\text{Aided maneuvers} = B + D + F = 24$$

$$\text{Disabled patient-handling operations} = A + B + D + E + F = 44$$

### 7.2.4 DESCRIPTION OF OPERATOR TRAINING

As was the case for the ward, in order to assess whether surgical unit staff had received adequate training for the specific risk in question, a number of aspects were looked into concerning the type of training delivered, namely,

- Was a theoretical and practical training course provided, or were operators merely given information?
- What was the duration of the course and time between last training and current risk assessment?
- What numbers of operators were involved in training courses/receiving information?

Another aspect to be considered is whether or not training effectiveness was assessed by means of written tests or assessments. Therefore, the data sheet is exactly the same as the one used for ward staff.

## 7.3   ON-SITE INSPECTION

If organizational aspects are reported during the interview with the head nurse of the surgical unit, the information concerning equipment and work environments is analyzed during the on-site visit.

### 7.3.1   ANALYSIS OF PATIENT-HANDLING DEVICES

The equipment routinely used by surgical unit staff to lift patients from "surface to surface" (e.g., operating table to stretcher, or ward bed to stretcher) generally includes

- Height-adjustable stretcher with sliding sheet or board
- Height-adjustable devices with horizontal sliding surface on castors (= mobilizer), mounted onto a wall separating the sterile area from the non-sterile area (= patient transfer unit)

The data sheet (Figure 7.1) then describes and briefly analyzes what equipment is present, how many devices, certain ergonomic features, if there is storage space for the equipment when not in use, and whether or not maintenance is adequate. Chapter 10 includes an in-depth analysis of the ergonomic requirements of this and other equipment and devices.

With regard to the analytical description of all the stretchers used in the surgical unit (stretcher "fleet"), as previously seen for wheelchairs in patient wards, an indication is given of any ergonomically significant aspects that may increase the frequency of manual patient handling and thus increase the risk of low-back biomechanical overload among patient-handling operators. The presence of these aspects is noted to define the *inadequacy score* for all stretchers of a particular type.

The total score (calculated by adding together the individual column scores), divided by the total number of stretchers, produces the **mean inadequacy score for the stretcher "fleet" (MSSTR)**. It is worth noting that this section describes both height-adjustable and nonadjustable stretchers.

### 7.3.2 ADEQUACY OF THE ENVIRONMENT

This section of the form describes and analyzes the operating room environment and relative fixtures and furnishings, with a specific focus on the operating table and/or anything that may hamper movement and/or the use of patient-handling equipment. Any inadequacies in the operating rooms or fixtures and furnishings will define the "mean inadequacy score for the environment" (MSENV) (Table 7.7).

**TABLE 7.7**

**Assessment of Stretchers and Environment/Furnishings in the Surgical Unit**

| Characteristics and Inadequacy Score for Stretchers | Score | Stretchers | | | | | |
|---|---|---|---|---|---|---|---|
| | | A (No.) | B (No.) | C (No.) | D (No.) | E (No.) | Total stretchers \|__\|__\| |
| Inefficient brakes | 1 | | | | | | |
| Not height adjustable | 2 | | | | | | |
| Side flaps | 2 | | | | | | |
| Needs to be partially lifted | 1 | | | | | | Total score |
| Column score (no. stretchers × sum of scores) | | | | | | | |
| Mean score MSSTR \|__\|, \|__\|__\| = Total score for stretchers / Total no. stretchers | | | | | | | |
| Characteristics and Inadequacy Score for Environment/Furnishings | Score | Environment/Furnishings | | | | | |
| | | Room A | Room B | Room C | Room D | Room E | Total rooms \|__\|__\| |
| Operating table with side rails | 2 | | | | | | |
| Nonremovable rails | 0.5 | | | | | | |
| Inadequate space for use of aids | 2 | | | | | | Total score |
| Column score (sum of scores) | | | | | | | |
| Mean score MSENV \|__\|, \|__\|__\| = Total score / Total no. operating rooms | | | | | | | |

**TABLE 7.8**

**Example of Equipment Description in a Surgical Unit**

| Equipment Description | No. | Lack of Essential Requirements | Lack of Adaptability to Patients or Environment | Lack of Maintenance |
|---|---|---|---|---|
| Sliding sheets | | Yes/No | Yes/No | Yes/No |
| Sliding boards | 2 | Yes/No | Yes/No | Yes/No |
| Other | | Yes/No | Yes/No | Yes/No |

### 7.3.3   EXAMPLE OF ON-SITE INSPECTION: DESCRIPTION OF EQUIPMENT AND ENVIRONMENT/FURNISHINGS

Following the on-site visit to the surgical unit, the following observations were made:

- Every OR was equipped with a sliding board.
- There were four OR stretchers—none height adjustable.
- OR "A" was too small to accommodate patient-handling aids; OR "B" had an operating table with side rails.

These observations are entered into Tables 7.8 and 7.9, which reproduce specific sections of the data collection sheet:

## 7.4   ESTIMATED RISK EXPOSURE IN SURGICAL UNIT AND SUMMARY OF RISK FACTORS

This analysis and description of risk factors in the surgical unit is an efficient way to estimate the level of manual patient-handling risk. The types of manual patient-handling activities performed in surgical units entails a high level of biomechanical risk (Waters, Nelson, and Proctor 2008; Chapter 3 in this book) and involves completely lifting patients who more often than not are under anesthesia. The "geometry" of the lifting action is such that the weight of the patient has to be supported horizontally over a considerable distance, generating an unacceptable degree of biomechanical overload.

The main factors based on which specific risk is estimated in the surgical unit (Figure 7.2) are the presence of procedures requiring patients to be lifted and the possibility of using lifting aids for such tasks. With regard to surgeries requiring

**TABLE 7.9**

**Example of Stretchers and Environment/Furnishings Description in a Surgical Unit**

| Characteristics and Inadequacy Score for Stretchers | Score | Stretchers | | | | | |
|---|---|---|---|---|---|---|---|
| | | A | B | C | D | E | |
| | | No. 2 | No. 2 | No. | No. | No. | |
| Inefficient brakes | 1 | | | | | | Total stretchers |
| Not height adjustable | 2 | x | x | | | | |_|_4_| |
| Side flaps | 2 | | | | | | |
| Needs to be partially lifted | 1 | | | | | | Total score |
| Column score (no. of stretchers × sum of scores) | | 4 | 4 | | | | 8 |

Mean score MSSTR 8/4 = 2 = $\dfrac{\text{Total score for stretchers}}{\text{Total no. stretchers}}$

| Characteristics and Inadequacy Score for Environment/ Furnishings | Score | Environment/Furnishings | | | | | |
|---|---|---|---|---|---|---|---|
| | | Room A | Room B | Room C | Room D | Room E | |
| | | 1 | 1 | | | | |
| Operating table with side rails | 2 | | x | | | | Total rooms |
| Nonremovable rails | 0.5 | | | | | | |_|_2_| |
| Inadequate space for use of aids | 2 | x | | | | | Total score |
| Column score (sum of scores) | | 2 | 2 | | | | 4 |

Mean score MSENV 4/2 = 2 = $\dfrac{\text{Total score}}{\text{Total no. operating rooms}}$

patients to be lifted (= NS), this factor is a necessary but not a sufficient factor, per se, for defining a specific risk, where the level depends on the percentage of aided patient lifting tasks. The possibility of moving patients using equipment determines the exposure level, as shown in Table 7.10.

In the overall grouping represented by the different surgical units, risk factors are listed based on the frequency of patient-handling tasks per individual operator (= F) expressed as the sum total of patient-handling operations multiplied by the number

**TABLE 7.10**
**Estimated Risk Exposure in Surgical Unit**

| Exposure Level | Risk Exposure Level | Frequency of Patient-Handling Tasks (F) |
|---|---|---|
| Absent | NS = 0 | |
| Negligible | NS ≠ 0, and AMPER ≥ 90% | \|__\|__\|,\|__\| F |
| High | NS ≠ 0, AND AMPER < 90% | \|__\|__\|,\|__\| F |

**TABLE 7.11**
**Summary of Equipment Assessment**

| Equipment Factor | | |
|---|---|---|
| Percentage of Aided Patient-Handling Tasks Observed \|__\|__\| AMPER | | Ergonomic Inadequacy |
| Equipment seldom used (AMPER < 50%) | | High |
| Equipment sometimes used (AMPER < 90% but ≥ 50%) | | Medium |
| Equipment adequately used (AMPER ≥ 90%) | | Negligible |

of operators performing these tasks daily (as shown in example 3B of Chapter 12). Therefore, the risk level plus the frequency of patient handling define priority actions for remediating risk within the surgical units analyzed.

The other factors (environment, stretchers, and training) steer the remediation process and lead to the attribution of mean scores for nonergonomic situations, corresponding to a value expressed using the traffic light system to indicate the level of criticality requiring action.

### 7.4.1 Equipment Factor

The analysis considers the percentage of aided patient-handling maneuvers routinely performed and broken down into three "ergonomic inadequacy" categories, according to criteria about percentage of aided patient-handling activities (Table 7.11).

### 7.4.2 Stretcher Factor, Environment Factor

To define the inadequacy of each individual factor, the same rationale used and described in the ward is also used in the surgical unit—that is, the relative "mean inadequacy score" (MSSTR – MSENV) calculated on the data sheet and broken down into three equidistant ranges (low, medium and high ergonomic inadequacy, respectively).

### 7.4.3 TRAINING FACTOR

This serves to describe the characteristics of the specific training/information delivered to surgical unit staff in terms of

- Adequacy of content
- Percentage of staff members trained
- Adequacy of period from training course (not too long a period)

The ergonomic inadequacy estimation associated with the training factor is according to the criteria in Figure 7.3.

FONDAZIONE IRCCS CA' GRANDA
Ospedale Maggiore Policlinico
CLINICA DEL LAVORO – MILAN (ITALY)
ERGONOMICS SECTION

## DATA COLLECTION SHEET: RISK ASSESSMENT FOR MANUAL PATIENT HANDLING RISK IN <u>SURGICAL UNIT</u>

HOSPITAL:_____SURGICAL UNIT :_____DATE: _____

### 1. INTERVIEW

| TOTAL OPERATORS (staff performing manual patient handling tasks): | | | |
|---|---|---|---|
| Nurses/scrub nurses: | Nurses aides: | | Other: |
| Surgical Unit daily working hours: | | Working days: | |
| OPERATORS (staff performing manual patient handling tasks) PRESENT OVER 24 HOURS: | | | |
| Nurses/scrub nurses: | Nurses aides: | | Other: |
| Indicate the number of operators (**Op**) as the sum of operators (in all job categories) involved in patient handling and present over 24 hours | | | **Op** |

### QUANTIFICATION OF SURGICAL PROCEDURES

| |
|---|
| Nr. of ORs _____ Nr. of procedures/year _____ Average nr. procedures/day \|_\|_\| |
| • Average nr. of procedures/day under general anesthesia (**GA**): _____ \|_\|_\| |
| • Average nr. of procedures/day under local anesthesia : _____ |
|     Average nr. of procedures/day **not requiring** total/partial patient lifting _____ |
|     Average nr. of procedures/day **requiring** total/partial patient lifting (**LA**) _____ |

| Nr. of procedures requiring patient handling (GA + LA) | **NS** |
|---|---|

### DISABLED PATIENT HANDLING OPERATIONS

| Type of anesthesia | GA = Nr_____ | | GA = Nr_____ | |
|---|---|---|---|---|
| ☐bed/stretcher | ☐manual | ☐aided | ☐manual | ☐aided |
| ☐stretcher/ operating table | ☐manual | ☐aided | ☐manual | ☐aided |
| ☐operating table / stretcher | ☐manual | ☐aided | ☐manual | ☐aided |
| ☐stretcher / bed | ☐manual | ☐aided | ☐manual | ☐aided |
| ☐stretcher / stretcher | ☐manual | ☐aided | ☐manual | ☐aided |
| ☐prone / supine | ☐manual | ☐aided | ☐manual | ☐aided |
| ☐supine / prone | ☐manual | ☐aided | ☐manual | ☐aided |
| Column score (Nr of maneuvers x Nr GA) | ____ A | ____ B | ____ C | ____ D |
| Type of anesthesia | LA = Nr_____ | | LA = Nr_____ | |
| ☐bed / stretcher | ☐manual | ☐aided | ☐manual | ☐aided |
| ☐stretcher / operating table | ☐manual | ☐aided | ☐manual | ☐aided |
| ☐operating table / stretcher | ☐manual | ☐aided | ☐manual | ☐aided |
| ☐stretcher / bed | ☐manual | ☐aided | ☐manual | ☐aided |
| ☐stretcher / stretcher | ☐manual | ☐aided | ☐manual | ☐aided |
| ☐prone / supine | ☐manual | ☐aided | ☐manual | ☐aided |
| ☐supine / prone | ☐manual | ☐aided | ☐manual | ☐aided |
| Column score (Nr of maneuvers x Nr LA) | ____ E | ____ F | ____ G | ____ H |

DISABLED PATIENT HANDLING OPERATIONS = A+B+C+D+E+F+G+H = \|__\|__\|

PERCENTAGE OF AIDED MANEUVERS: Sum of scores: $\dfrac{B + D + F + H}{A+B+C+D+E+F+G+H} \times 100$  => \|__\|__\| AMPER

**FIGURE 7.1**   Surgical data sheet. *(continued)*

FONDAZIONE IRCCS CA' GRANDA
OSPEDALE MAGGIORE POLICLINICO
CLINICA DEL LAVORO – MILAN (ITALY)
ERGONOMICS SECTION

| OPERATOR EDUCATION AND TRAINING | | | | | |
|---|---|---|---|---|---|
| **EDUCATION AND TRAINING** | | | **INFORMATION** | | |
| Attended theoretical/practical course | ☐ YES | ☐ NO | Training only on how to use equipment | ☐ YES | ☐ NO |
| if YES, how many months ago? and how many hours/operator | Months _____ hours _____ | | Only provided brochures on MPH | ☐ YES | ☐ NO |
| if YES, how many operators? | | | if YES, how many operators? | | |
| Was EFFECTIVENESS measured and documented in writing? | | | ☐ YES | ☐ NO | |

### 2. ON-SITE INSPECTION

**DESCRIPTION OF DISABLED PATIENT LIFTING/TRANSFERRING EQUIPMENT**

| EQUIPMENT DESCRIPTION | | Nr | Lack of essential requirements | Lack of adaptability to patients or environment | Lack of maintenance |
|---|---|---|---|---|---|
| **"MOBILIZER"**= power driven height-adjustable device with sliding board for patient transfer | | | YES   NO | YES   NO | YES   NO |
| **"PATIENT TRANSFER UNIT"** = **wallmounted** power driven height-adjustable device separating operating room from preoperating holding area, featuring sliding board for patient transfer | | | YES   NO | YES   NO | YES   NO |
| Adjustable **STRETCHER** type: | | | YES   NO | YES   NO | YES   NO |
| Adjustable **STRETCHER** type: | | | YES   NO | YES   NO | YES   NO |

| EQUIPMENT DESCRIPTION | | Nr | Lack of essential requirements | Lack of adaptability to patients or environment | Lack of maintenance |
|---|---|---|---|---|---|
| SLIDING SHEETS: | | | YES   NO | YES   NO | YES   NO |
| SLIDING BOARDS: | | | YES   NO | YES   NO | YES   NO |
| OTHER: | | | YES   NO | YES   NO | YES   NO |

| |
|---|
| STORAGE ROOM/SPACE FOR EQUIPMENT WHEN NOT IN USE ☐ NO   ☐ YES   ☐ if YES, m² |
| ☐ MAINTENANCE ADEQUATE - IF NOT, specify_____ |

**FIGURE 7.1** *(continued)*   Surgical data sheet. *(continued)*

**DESCRIPTION OF STRETCHERS AND ENVIRONMENT/FURNISHINGS**

| CHARACTERISTICS AND INADEQUACY SCORE FOR STRETCHERS | SCORE | STRETCHERS | | | | | Total Nr stretchers |
|---|---|---|---|---|---|---|---|
| | | A | B | C | D | E | |_|_| |
| | | Nr | Nr | Nr | Nr | Nr | |
| Inefficient brakes | 1 | | | | | | |
| Not height-adjustable | 2 | | | | | | |
| Side flaps | 2 | | | | | | Total score |
| Needs to be partially lifted | 1 | | | | | | |
| **Column score (Nr of stretchers × sum of scores)** | | | | | | | |

Mean score **MSSTR** |__|. |__|__| = $\dfrac{\text{Total score for stretchers}}{\text{Total Nr stretchers}}$

| CHARACTERISTICS AND INADEQUACY SCORE FOR ENVIRONMENT/FURNISHINGS | SCORE | Environment/furnishings | | | | | Total Nr rooms |
|---|---|---|---|---|---|---|---|
| | | Room A | Room B | Room C | Room D | Room E | |_|_| |
| Operating table with side rails | 2 | | | | | | |
| Non-removable rails | 0.5 | | | | | | |
| Inadequate space for use of aids | 2 | | | | | | Total Score |
| **Column score (sum of scores)** | | | | | | | |

Mean score **MSENV** |__|, |__|__| = $\dfrac{\text{Total score}}{\text{Total Nr operating rooms}}$

**NOTE:**

_____

_____

**FIGURE 7.1** *(continued)*   Surgical data sheet.

FONDAZIONE IRCCS CA' GRANDA
OSPEDALE MAGGIORE POLICLINICO
CLINICA DEL LAVORO   MILAN ITALY
ERGONOMICS SECTION

## SURGICAL UNITS: HOW TO ASSIGN "VALUES" TO RISK FACTORS

N.B. If a surgical unit is comprised of several operating rooms and staff does not rotate between them all, a separate form must be completed for each room.

The average number of procedures/day must be calculated by dividing the total number of procedures performed during the whole year by the total number of days that the surgical unit is used.

(OP) indicates the sum total patient handling operators present over 24 hours.

(if any operators are unauthorized or exempted from lifting activities and are not assigned to MPH tasks, do not count them as OPs).

Frequency of patient handling tasks (F) indicates the ratio of patient handling maneuvers/tasks sum (both manual and aided) to the number of operators present over 24 hours (Op).

### EQUIPMENT FACTOR

THE ADEQUACY OF LIFTING DEVICES IS DETERMINED BASED ON THE PERCENTAGE OF AIDED PATIENT MANEUVERS (**AMPER**)

if **AMPER** is below 50% = **INADEQUACY HIGH**
if **AMPER** is below 90% but above (or equal to) 50% = **MEDIUM INADEQUACY**
if **AMPER** is at least 90% = **INADEQUACY NEGLIGIBLE**

### STRETCHERS FACTOR

To define the wheelchairs/stretchers factor, it is necessary to assess the "MEAN INADEQUACY SCORE" reported in the data collection sheet (**MSSTR**) as shown in the following table:

| STRETCHERS FACTOR (SF) | | | |
|---|---|---|---|
| Mean qualitative score observed (MSSTR) | 0.0 – 2.00 | 2.01 - 4.00 | 4.01- 6 |
| ERGONOMIC INADEQUACY | NEGLIGIBLE | MEDIUM | HIGH |

### ENVIRONMENT FACTOR

To define the environment factor, it is necessary to assess the "mean inadequacy score" for the environment (**MSENV**) reported in the data collection sheet is divided into three equidistant ranges (i.e. low, medium and high inadequacy), as shown in the following table.

| MSENV | 0-1.5 | 1.51-3 | 3.01- 4.5 |
|---|---|---|---|
| ERGONOMIC INADEQUACY | NEGLIGIBLE | MEDIUM | HIGH |

### TRAINING FACTOR

As in the case of wards, the various aspects of training are described.

| Characteristics observed | ERGONOMIC INADEQUACY |
|---|---|
| Adequate training course held no more than 2 years before the risk assessment, attended by 75% of surgical unit staff | NEGLIGIBLE |
| Training course held more than 2 years before the risk assessment, attended by 75% of surgical unit staff, and tested for effectiveness | NEGLIGIBLE |
| Adequate training course held no more than 2 years before the risk assessment, attended by between 50% and 75% of surgical unit staff | MEDIUM |
| Only explanations (or specific brochures) given to 90% of surgical unit staff, followed by effectiveness testing | MEDIUM |
| TRAINING NOT DELIVERED OR NOT COMPLIANT WITH AFOREMENTIONED CONDITIONS | HIGH |

An **adequate training course** is defined as a theoretical/practical course of at least 6 hours with some of the practical training devoted to the use of lifting aids.

TO DETERMINE EXPOSURE LEVEL, ENTER DATA INTO THE SOFTWARE

**FIGURE 7.2** Surgical unit legend.

FONDAZIONE IRCCS CA' GRANDA
OSPEDALE MAGGIORE POLICLINICO
CLINICA DEL LAVORO – MILAN, ITALY
ERGONOMICS SECTION

SUMMARY FOR SURGICAL UNIT          Date _____

| Hospital | Surgical Unit | Code |
|---|---|---|
| Number of procedures/day requiring patient handling (GA + LA) | | \|__\|__\| NS |
| OPERATORS (staff performing manual patient handling tasks) PRESENT OVER 24 HOURS: | | \|__\|__\|Op |

| FREQUENCY OF PATIENT HANDLING TASKS | A+B+C+D+E+F+G+H / \|__\|__\|Op | \|__\|__\|,\|__\| F |
|---|---|---|

| EQUIPMENT FACTOR | |
|---|---|
| PERCENTAGE OF AIDED PATIENT HANDLING TASKS observed \|__\|__\| AMPER | ERGONOMIC INADEQUACY |
| EQUIPMENT SELDOM USED (AMPER below 50%) | HIGH |
| EQUIPMENT SOMETIMES USED (AMPER below 90% but above (or equal to) 50%) | MEDIUM |
| EQUIPMENT ADEQUATELY USED (AMPER ≥ 90%) | NEGLIGIBLE |

| EXPOSURE LEVEL | TICK EXPOSURE LEVEL | FREQUENCY OF PATIENT HANDLING TASKS (F) |
|---|---|---|
| ABSENT | NS = 0 | |
| NEGLIGIBLE | NS ≠ 0, AND AMPER ≥ 90% | \|__\|__\|,\|__\| F |
| HIGH | NS ≠ 0, AND AMPER < 90% | \|__\|__\|,\|__\| F |

### OTHER IMPORTANT ASPECTS WITH REGARD TO RISK REDUCTION

| STRETCHERS FACTOR | | | | (MSSTR) observed: |
|---|---|---|---|---|
| Mean qualitative score observed (MSSTR) | 0.0 – 2.00 | 2.01 - 4.00 | 4.01- 6 | \|__\|__\|,\|__\| |
| ERGONOMIC INADEQUACY | NEGLIGIBLE | MEDIUM | HIGH | |

| ENVIRONMENT FACTOR | | | | (MSENV) observed: |
|---|---|---|---|---|
| | | | | \|__\|__\|,\|__\| |
| Mean score non-ergonomic conditions observed (MSENV) | 0.0 -1.5 | 1.51 - 3 | 3.01- 4.5 | |
| ERGONOMIC INADEQUACY | NEGLIGIBLE | MEDIUM | HIGH | |

| TRAINING FACTOR | | | |
|---|---|---|---|
| Type of training | Adequate | Partly adequate | Completely inadequate |
| ERGONOMIC INADEQUACY | NEGLIGIBLE | MEDIUM | HIGH |

**FIGURE 7.3**   Summary for surgical unit.

# 8 Analysis of Outpatient Services and Day Hospital

## 8.1 INTRODUCTION

Hospitals have changed considerably over the past 10 years, leading them to focus increasingly on dealing with emergencies and acute pathologies (where most patients are not mobile, due to the motor impairments determined by acute disease). At the same time, shorter hospital stays have led to greater reliance on outpatient services, day hospitals, and so on to provide patients with postcritical treatment and care. Consequently, ever larger numbers of disabled (D) patients are turning to outpatient services while the number of operators staffing them is often insufficient.

An unfavorable ratio between the number of noncooperative (NC) patients and the number of operators (OPs) performing manual patient-handling tasks (NC/OP) is a determining factor in defining specific risk.

Outpatient services commonly employ healthcare workers affected by pathologies that are incompatible with biomechanical overload (manual patient handling) or with night shifts. In addition, workers in this specific sector are often older and have been in their jobs longer (often also with long-standing risk exposure), making it very difficult to manage job suitability among a significant proportion of healthcare workers.

On the other hand, the available literature is not particularly helpful in defining risk analysis methodologies in outpatient facilities. The procedure is described in the following paragraphs for quantifying patient-handling risk in

- Outpatient services in different healthcare settings
- Day hospitals
- Radiology departments

A closer look at the data sheet for detecting risk in these facilities (Figure 8.1) reveals a general data collection format that is identical to the one presented in Chapter 5 of this book.

## 8.2 DATA COLLECTION SHEET: ORGANIZATIONAL ASPECTS

The general approach used for the department is the same and entails gathering information via an interview with the head of the department and an on-site visit. In the interview, after explaining why the information is being collected, the head

**TABLE 8.1**

**Description of Staff Engaged in MPH Tasks**

| No. operators performing MPH: indicate the total number of operators per job category | | | |
|---|---|---|---|
| Nurses: | Orderlies (ASA/OTA/OSA): | Nurses aides: | Other: |

**No. operators performing MPH tasks over three shifts:** indicate number of operators on duty per shift

| Shift | Morning | Afternoon | Night |
|---|---|---|---|
| Shift hours: (from 00:00 to 00:00) | From_____to_____ | From_____to_____ | From_____to_____ |
| No. operators over entire shift | | | |
| (A) Total operators over entire shift = | | | |

**No. part-time operators:** indicate the exact number of hours worked and calculate them as unit fractions (in relation to the overall duration of the shift)

| No. Part-Time Operators Present | Hours Worked in Shift (from 00:00 to 00:00) | Unit Fraction | Unit Fraction by No. Operators |
|---|---|---|---|
| | From_____to_____ | | |
| | From_____to_____ | | |
| (B) Total operators (as unit fractions) present by shift duration = | | | |
| **Total no. operators engaged in MPH over 24 hours (OPs):** add the total number of operators present over the entire shift (A) to the total number of part-time operators (B) | | | **Op** |

of the department (or an "experienced" staff member) is asked to describe the relevant organizational aspects (i.e., details concerning the staff involved in patient-handling activities), using the data collection sheet in Table 8.1.

The usual three-shift arrangement is used for this calculation of the number of operators on hand, although certain radiology or day hospital facilities run on two or three shifts; if the staff work just one shift, then only one shift will be described.

The evaluation of staff training with respect to specific manual patient-handling (MPH) risk is the same in this setting as for the standard ward, with the same aspects being analyzed:

- Theoretical–practical training course or information only
- Course duration and time between last training course and current risk assessment
- Number of operators receiving training/information

The analysis also looks at whether or not the effectiveness of the training has been measured, by examining any written evidence of such an assessment.

## 8.2.1 QUANTIFICATION OF THE AVERAGE NUMBER OF NC AND PC PATIENTS

In outpatient facilities, the number of disabled patients is analyzed versus the average number of daily visits to the service. As in the case of hospital wards, the daily number of dependent patients will be broken down into NC (noncooperative) and PC (partially cooperative), based on how they are lifted by workers (i.e., either totally lifted or partially lifted). If these numbers are difficult to calculate, the following method should be used:

1. Define the average number of patient visits/day to the service.
2. Define the percentage of disabled patients versus the total number of patient visits (this figure is usually easy to find).
3. Fill in Table 8.2, which reproduces the relevant section of the data collection sheet.

If this is too difficult, or if a more objective evaluation is preferred, it is also possible to keep a daily patient diary, entering each type of patient visit, and then subsequently calculating the mean value for the specific setting. The number of days in which to carry out this analysis will be determined based on the information provided by the head of the service. The analysis may be carried out using a similar form to the one shown in Table 8.3.

## 8.2.2 DESCRIPTION AND QUANTIFICATION OF DISABLED PATIENT-HANDLING TASKS

The main difference between outpatient services and wards lies in the greater amount of detail regarding the patient-handling tasks: In outpatient settings, most

---

**TABLE 8.2**

**Quantification of Disabled Patients in the Outpatient Service**

**Average number/day of visits by disabled patients (D)**

(If it is difficult to quantify the average number, indicate the percentage (_%) versus the number of visits/day.)

_____ D

**By whom**

Indicate the average number of totally noncooperative patients (NC) visiting the service

NC = patient who needs to be completely lifted for transfer and repositioning operations (TL)

_____ NC

Indicate the average number of partially cooperative patients (PC) visiting the service

PC = patient who only needs to be partially lifted (PL)

_____ PC

**TABLE 8.3**

**Patient Visit Diary**

| Date | Type of Patient Visit | | |
|---|---|---|---|
| | Self-Sufficient Patient | PC | NC |
| 1. Visit | | | |
| 2. Visit | | | |
| 3. Visit | | | |
| 4. Visit | | | |
| 5. Visit | | | |
| 6. Visit | | | |
| 7. Visit | | | |
| 8. Visit | | | |
| 9. Visit | | | |
| 10. Visit | | | |

patient-handling activities consist of transferring patients from the vehicle or wheelchair in which they arrive at the service to the examination bed or radiology table, and vice versa. It is therefore relatively easy to identify the different types of transfers (e.g., between stretcher and examination bed, or between wheelchair and examination bed) and the average number of transfers per type—always, of course, in relation to the number of disabled patients.

Therefore, all handling operations (either manual or aided) featuring total lifting (TL) or partial lifting (PL) will be described. The sum of total lifting operations (both aided and manual) = Ab + B + C + D = sum TL.

$$\text{Percentage of aided total lifts (\%ATL)} = \text{sum of scores: } \frac{B+D}{\text{SUM TL}}$$

The sum of partial lifting operations (both aided and manual) = E + F + G + H + X + Z + W + K = sum PL.

$$\text{Percentage of aided partial lifts (\%APL)} = \text{sum of scores: } \frac{F+H+Z+K}{\text{SUM PL}}$$

### 8.2.2.1 Example of Quantification of Lifting Maneuvers

The example shown in Table 8.4 indicates the number of patient-handling tasks for an outpatient service that routinely receives ten NC patients (i.e., patients requiring

## TABLE 8.4
## Example 1: Description and Quantification of Lifting Maneuvers

| 1. Total Lifts (TL) | NC = No. __10__ | | NC = No. ____ | |
|---|---|---|---|---|
| Stretcher/exam bed | [ ] Manual | [X] aided | [ ] Manual | [ ] Aided |
| Wheelchair/exam bed | [ ] Manual | [ ] Aided | [ ] Manual | [ ] Aided |
| Ward bed/exam bed | [ ] Manual | [ ] Aided | [ ] Manual | [ ] Aided |
| Exam bed/stretcher | [ ] Manual | [X] aided | [ ] Manual | [ ] Aided |
| Exam bed/wheelchair | [ ] Manual | [ ] Aided | [ ] Manual | [ ] Aided |
| Exam bed/ward bed | [ ] Manual | [ ] Aided | [ ] Manual | [ ] Aided |
| Other | [ ] Manual | [ ] Aided | [ ] Manual | [ ] Aided |
| Total column score (no. maneuvers per no. patients | A ____ | B __20__ | C ____ | D ____ |

| 2. Partial Lifts (PL) for PC Patients | PC = No. __5__ | | PC = No. ____ | |
|---|---|---|---|---|
| Stretcher/exam bed | [ ] Manual | [ ] Aided | [ ] Manual | [ ] Aided |
| Wheelchair/exam bed | [X] manual | [ ] Aided | [ ] Manual | [ ] Aided |
| Ward bed/exam bed | [ ] Manual | [ ] Aided | [ ] Manual | [ ] Aided |
| Exam bed/stretcher | [ ] Manual | [ ] Aided | [ ] Manual | [ ] Aided |
| Exam bed/wheelchair | [X] manual | [ ] Aided | [ ] Manual | [ ] Aided |
| Exam bed/ward bed | [ ] Manual | [ ] Aided | [ ] Manual | [ ] Aided |
| Other | [ ] Manual | [ ] Aided | [ ] Manual | [ ] Aided |
| Total column score (no. maneuvers per no. patients | E __10__ | F ____ | G ____ | H ____ |

| 3. Partial Lifts (PL) per NC Patients | NC = No. __10__ | | NC = No. ____ | |
|---|---|---|---|---|
| Turning over | [X] manual | [ ] Aided | [ ] Manual | [ ] Aided |
| Lifting patient's upper body | [X] manual | [ ] Aided | [ ] Manual | [ ] Aided |
| Other | [ ] Manual | [ ] Aided | [ ] Manual | [ ] Aided |
| Other | [ ] Manual | [ ] Aided | [ ] Manual | [ ] Aided |
| Total column score (no. maneuvers per no. patients | X __20__ | Z ____ | W ____ | K ____ |

total lifting) and five PC patients (i.e., who require only partial lifting) per day. The NC patients arrive on stretchers and are transferred using sliding boards, while the PC patients arrive in wheelchairs and are lifted manually (sections 1 and 2 of the table). On the examination bed, NC patients are also pulled up to a sitting position or turned over—that is, partially lifted (section 3 of the table).

The description in Table 8.5 of handling tasks will also serve to quantify the percentage of aided maneuvers, broken down into aided *total* lifts (ATLPER) and aided *partial* lifts (APLPER).

During the on-site visit, it will be necessary to verify the organizational aspects described during the interview and also to check the ergonomic adequacy of both the equipment used for patient-handling activities and the environment in which the equipment is employed.

**TABLE 8.5**
**Percentage of Aided Maneuvers in Example 1**

| | |
|---|---|
| Sum of total lifting operations (both aided and manual) = A + B + C +D = \|_2\|0_\| sum TL | |
| Percentage of aided total lifts (% TL) = sum of scores: 20/20 | ATLPER = 100% |
| Sum of partial lifting operations (both aided and manual) = E + F + G + H + X + Z + W + K = \|_3\|0\| sum PL | |
| Percentage of aided partial lifts (% PL) = sum of scores: 0/30 | APLPER = 0% |

## 8.3  DESCRIPTION OF PATIENT-HANDLING ASSISTIVE DEVICES

The assistive devices used to lift/transfer patients (generally used in the service) include

- Hoist or sliding sheet or rollers (to help transfer the patient from wheelchair to examination bed and vice versa)
- Height-adjustable stretcher + sliding sheet/board (to help transfer the patient from stretcher to examination bed and vice versa
- Sliding sheets (to help turn/reposition the patient on the examination bed)
- Height-adjustable examination bed (to avoid uneven surfaces)

The devices are thus carefully examined as in the case of the ward, with a focus on their ergonomic adequacy (Chapter 10 of this book), number, and actual utilization (Table 8.6).

Next, the characteristics of the patient transportation equipment are described and quantified (i.e., wheelchairs and stretchers). The "mean inadequacy score" for wheelchairs is calculated as for the ward, with the addition here of stretchers, using the same method of assigning a score to the inadequacy characteristics (Table 8.7).

In outpatient services, stretchers must be able to be adjusted so as to eliminate differences in height between the stretcher and the examination bed or other surface, must not have side flaps, and must have upper sections that do not require manual lifting.

**TABLE 8.6**
**Assessment of Ergonomic Requirements**

## TABLE 8.7
## Method for Describing Different Types of Stretchers and Wheelchairs

| Characteristics and Inadequacy Score for Stretchers | Score | Types of Stretchers | | | | | |
|---|---|---|---|---|---|---|---|
| | | A (No.) | B (No.) | C (No.) | D (No.) | E (No.) | |
| Malfunctioning brakes | 1 | | | | | | Total stretchers ⌊_⌊_⌋ |
| Not height adjustable | 2 | | | | | | |
| Side flaps | 2 | | | | | | |
| Needs to be partially lifted | 1 | | | | | | Total score |
| **Column score (no. stretchers × sum of scores)** | | | | | | | |
| Characteristics and Inadequacy Score for Wheelchair | Score | Types of Wheelchairs | | | | | |
| | | A (No.) | B (No.) | C (No.) | D (No.) | E (No.) | |
| Malfunctioning brakes | 1 | | | | | | Total wheelchairs ⌊_⌊_⌋ |
| Nonremovable armrests | 1 | | | | | | |
| Cumbersome backrest | 1 | | | | | | |
| Width exceeding 70 cm | 1 | | | | | | Total score |
| **Column score (no. wheelchairs × sum of scores)** | | | | | | | |

## TABLE 8.8
## Summary and Quantification of Inadequacies for Stretchers and Wheelchairs

Mean score (**MSSTR**) = total stretchers score/total no. stretchers

Mean score (**MSWh**) = total wheelchairs score/total no. wheelchairs

**Mean wheelchairs and stretchers inadequacy score = (MSSTR + MSWh)**

The purpose of this description is to calculate the mean inadequacy score for wheelchairs and stretchers as a single unit, which is the sum of the mean inadequacy scores obtained separately for wheelchairs and stretchers, as shown in Table 8.8.

## 8.4 ADEQUACY OF ENVIRONMENT AND FURNISHINGS

During the on-site visit, the environment in which manual patient-handling activities take place, the furnishings, and the layout of the facility are analyzed. The structural characteristics of the facility are evaluated, as in the case of the ward, allowing, of

**TABLE 8.9**
**Description of Outpatient Environment/Furnishings**

### Characteristics of Examination Rooms

Free space inadequate for use of aids

Exam bed not height adjustable

Side flaps–exam bed

Part of exam bed needs to be raised manually

Patient armchair height less than 50 cm

Door width < 85 cm

### Characteristics of DH Bedrooms

Space between beds or between bed and wall less than 90 cm

Space between foot of bed and wall less than 120 cm

Unsuitable bed that needs to be partially lifted

Space between bed and floor less than 15 cm

Patient armchair height less than 50 cm

course, for differences between outpatient examination rooms and any patient bedrooms that may be present in day hospitals, for instance, and excluding bathrooms from the mean environment score, since these are seldom used by disabled patients.

More specifically, nonergonomic conditions are evidenced (Table 8.9) based on a score that is also used to define the total inadequacy score for DH rooms and bedrooms. The overall "inadequacy" of the environment is expressed by a mean score (MSENV), calculated as in the data collection sheet for the ward.

## 8.5  ASSESSMENT OF MPH RISK EXPOSURE IN OUTPATIENT SERVICES: SUMMARY AND CRITERIA FOR ESTIMATING EXPOSURE

As was done for the surgical block, the analysis of manual patient-handling risk factors in Outpatient Services, while not leading to the definition of a short index, does provide a classification of exposure levels as well as useful recommendations for overcoming any peak risks and deciding what factors need to be corrected so as to achieve the best possible results.

Once again, the criteria based on which risk is estimated will include

- Patients needing to be moved
- The percentage of patient handling aided

The main factor (i.e., visits by disabled [D] patients to the outpatient service) is a necessary, although per se not sufficient, condition for defining specific risk, where the risk level is determined by the presence of patients to be fully or partially lifted, the percentage of aided patient-handling tasks (ATLPER and APLPER); these aspects indirectly indicate the inadequacy of the patient-handling aids utilized (for total lifting and minor aids).

The summary of outpatient/day hospital services (Figure 8.3) contains all the risk factors described and listed in the following:

**Handling frequency (F):** ratio of the sum of patient-handling maneuvers/tasks (manual and aided) to the total number of operators present over the 24-hour period (OPs). The corrective action plan for reducing risk levels will be prioritized based on the exposure level observed, among which frequency (F) of patient handling per operator (indicating the intensity of patient-handling activities) represents an important aspect.

**Equipment factor for total lifting:** indicates the adequacy of the lifting aids based on the percentage of aided total lifts (ATLPER). Table 8.10 summarizes the rationale, also used previously, for the breakdown into three "inadequacy" levels.

**Equipment factor for partial lifting:** The adequacy of patient lifting aids is assessed based on the percentage of aided partial lifting (ATTPER PL), broken down into two levels, as shown in Figure 8.2. The other risk factors described relate to transportation equipment (stretchers and wheelchairs), environment, and training; here, the evaluation leads to the allocation of mean nonergonomic scores corresponding to a value that indicates the level of ergonomic inadequacy (i.e., criticalities requiring corrective action).

**Wheelchair/stretcher factor:** to assess the risk factor generated by the possible inadequacy of stretchers and wheelchairs, the relative mean inadequacy score is obtained from the data collection sheet (MSSTR + MSWh), and then the degree of wheelchair/stretcher inadequacy is identified as shown in Figure 8.3.

**Environment factor:** to evaluate the environment factor, the mean inadequacy score for the environment (MSENV) is calculated on the data sheet and broken down into three equidistant ranges (low, medium, and high ergonomic inadequacy).

**TABLE 8.10**

**Equipment Factor for Total Lifting: Inadequacy Levels**

Total Lifting Devices Factor

| Percentage of Total Aided Lifts \|__\|__\| ATLPER | Ergonomic Inadequacy |
| --- | --- |
| Equipment seldom used (ATLPER < 50%) | High |
| Equipment sometimes used (ATLPER < 90% but ≥ 50%) | Medium |
| Equipment adequately used (ATLPER ≥ 90%) | Negligible |

**TABLE 8.11**
**Risk Exposure Levels in Outpatient Services**

| Exposure Level | Risk Exposure Level | Frequency of Patient-Handling Tasks (F) |
|---|---|---|
| Absent | D = 0 | |
| Negligible | ATLPER ≥ 90% and APLPER ≥ 90% | \|__\|__\|,\|__\| F |
| Medium | ATLPER ≥ 90% and APLPER < 90% | \|__\|__\|,\|__\| F |
| High | <90% | \|__\|__\|,\|__\| F |

**Training factor:** training for staff exposed to risk is evaluated using the same criteria as those described previously for wards and surgical blocks, in order to define the ergonomic inadequacy of the training factor. Table 8.11 shows exposure levels to the specific risk and the criteria employed for determining them.

FONDAZIONE IRCCS CA' GRANDA
OSPEDALE MAGGIORE POLICLINICO
CLINICA DEL LAVORO – MILAN ITALY
ERGONOMICS SECTION

**DATA COLLECTION SHEET: RISK ASSESSMENT FOR MANUAL PATIENT HANDLING IN**
**OUTPATIENT/DAY HOSPITAL SERVICES**

HOSPITAL: _____        SERVICE : _____        Code _____ Date _____

NUMBER OF VISITS/DAY: _____ OPENING HOURS: from _____ to _____ Days open _____

**1. INTERVIEW**

| **Nr OF OPERATORS PERFORMING MPH:** indicate the total number of operators per job category. | | | |
|---|---|---|---|
| nurses: | orderlies (ASA/OTA/OSA): | nurses aides: | other: |
| **Nr OPERATORS PERFORMING MPH TASKS OVER 3 SHIFTS:** indicate number of operators on duty per shift: | | | |
| SHIFT | **morning** | **afternoon** | **night** |
| Shift hours:<br>(from 00:00 to 00:00) | from_____to_____ | from_____to_____ | from_____to_____ |
| Nr of operators over entire shift | | | |
| **(A)** Total operators over entire shift = | | | |
| **Nr OF PART-TIME OPERATORS:** indicate the exact number of hours worked and calculate them as unit fractions (in relation to the overall duration of the shift). | | | |
| Nr of part-time operators present | Hours worked in shift:<br>(from 00:00 to 00:00) | Unit fraction | (unit fraction by nr of operators) |
| | from_____to_____ | | |
| | from_____to_____ | | |
| **(B)** Total operators (as unit fractions) present by shift duration = | | | |
| **TOTAL Nr OF OPERATORS ENGAGED IN MPH OVER 24 HOURS (Op):** add the total number of operators present over the entire shift **(A)** to the total number of part-time operators **(B)** | | | **Op** |

| **AVERAGE NUMBER OF DISABLED PATIENTS (D):** |
|---|
| Average number/day of visits by disabled patients (D)     _____ D |
| (if it is difficult to quantify the average number, indicate the percentage _____%, versus the number of visits/day) |
| **of whom** |
| Indicate the average number of totally non cooperative patients (NC) visiting the service<br>NC= patient who needs to be completely lifted for transfer and repositioning operations (TL)     _____ NC |
| Indicate the average number of partially cooperative patients (PC) visiting the service.<br>PC= patient who only needs to be partially lifted (PL)     _____ PC |

**FIGURE 8.1**   Services sheet.

FONDAZIONE IRCCS CA' GRANDA
OSPEDALE MAGGIORE POLICLINICO
CLINICA DEL LAVORO – MILAN, ITALY,
ERGONOMICS SECTION

**DESCRIPTION OF MANUAL/AIDED PATIENT HANDLING OPERATIONS:**

| 1) Total lifts (TL) | NC = Nr _____ | | NC = Nr _____ | |
|---|---|---|---|---|
| ☐ stretcher/exam bed | ☐ manual | ☐ aided | ☐ manual | ☐ aided |
| ☐ wheelchair/exam bed | ☐ manual | ☐ aided | ☐ manual | ☐ aided |
| ☐ ward bed/exam bed | ☐ manual | ☐ aided | ☐ manual | ☐ aided |
| ☐ exam bed/ stretcher | ☐ manual | ☐ aided | ☐ manual | ☐ aided |
| ☐ exam bed/wheelchair | ☐ manual | ☐ aided | ☐ manual | ☐ aided |
| ☐ exam bed/ward bed | ☐ manual | ☐ aided | ☐ manual | ☐ aided |
| ☐ other | ☐ manual | ☐ aided | ☐ manual | ☐ aided |
| SCORE Column total (Nr of maneuvers × Nr of patients) | A___ | B___ | C___ | D___ |

SUM of total lifting operations (both aided and manual) = A+B+C+D = |__|__| SUM TL

Percentage of aided total lifts (%TL) = sum of scores: $\dfrac{B+D}{SUM\ TL} \times 100 =$ |__|__|

ATLPER

| 2) Partial lifts (PL) of PC patients | PC = Nr _____ | | PC = Nr _____ | |
|---|---|---|---|---|
| ☐ stretcher /exam bed | ☐ manual | ☐ aided | ☐ manual | ☐ aided |
| ☐ wheelchair/exam bed | ☐ manual | ☐ aided | ☐ manual | ☐ aided |
| ☐ ward bed/exam bed | ☐ manual | ☐ aided | ☐ manual | ☐ aided |
| ☐ exam bed/ stretcher | ☐ manual | ☐ aided | ☐ manual | ☐ aided |
| ☐ exam bed/wheelchair | ☐ manual | ☐ aided | ☐ manual | ☐ aided |
| ☐ exam bed/ward bed | ☐ manual | ☐ aided | ☐ manual | ☐ aided |
| ☐ other | ☐ manual | ☐ aided | ☐ manual | ☐ aided |
| Column total score (Nr of maneuvers × Nr of patients) | E___ | F___ | G___ | H___ |
| 3) Partial lifts (PL) of NC patients | NC = Nr _____ | | NC = Nr _____ | |
| ☐ turning over | ☐ manual | ☐ aided | ☐ manual | ☐ aided |
| ☐ pulling up in bed | ☐ manual | ☐ aided | ☐ manual | ☐ aided |
| ☐ other | ☐ manual | ☐ aided | ☐ manual | ☐ aided |
| ☐ other | ☐ manual | ☐ aided | ☐ manual | ☐ aided |
| Column total score (Nr of maneuvers × Nr of patients) | X___ | Z___ | W___ | K___ |

SUM of partial lifting operations (both aided and manual) = E+F+G+H+X+Z+W+K = |__|__| SUM PL

Percentage of aided partial lifts (%PL) = sum of scores: $\dfrac{F\ H\ Z\ K}{SUM\ PL} \times 100 =$ |__|__|

APLPER

**DISABLED PATIENT HANDLING OPERATIONS = SUM TL + SUM PL = |__|__|**

| OPERATOR TRAINING | | | | | | |
|---|---|---|---|---|---|---|
| **TRAINING** | | | **INFORMATION** | | | |
| Attended theoretical/practical course | ☐ YES | ☐ NO | Training only on how to use equipment | ☐ YES | | ☐ NO |
| if YES, how many months ago? and how many hours/operator | Months _____ hours _____ | | Only provided brochures on MPH | ☐ YES | | ☐ NO |
| if YES, how many operators? | | | if YES, how many operator? | | | |
| Was EFFECTIVENESS measured and documented in writing? | | ☐ YES | | | ☐ NO | |

**FIGURE 8.1** *(continued)* Services sheet. *(continued)*

FONDAZIONE IRCCS CA' GRANDA
OSPEDALE MAGGIORE POLICLINICO
CLINICA DEL LAVORO – MILAN (ITALY)
ERGONOMICS SECTION

2. <u>ON-SITE INSPECTION</u>

**EQUIPMENT FOR DISABLED PATIENT LIFTING/TRANSFERS**

| EQUIPMENT DESCRIPTION | | Nr | Lack of essential requirements | | Lack of adaptability to patients or environment | | Lack of maintenance | |
|---|---|---|---|---|---|---|---|---|
| LIFTING EQUIPMENT type: | | | YES      NO | | YES      NO | | YES      NO | |
| Adjustable EXAM BED: | | | YES      NO | | YES      NO | | YES      NO | |
| Adjustable STRETCHER: | | | YES      NO | | YES      NO | | YES      NO | |
| OTHER: | | | YES      NO | | YES      NO | | YES      NO | |

**OTHER AIDS (MINOR AIDS):**

| EQUIPMENT DESCRIPTION | | Nr | Lack of essential requirements | | Lack of adaptability to patients or environment | | Lack of maintenance | |
|---|---|---|---|---|---|---|---|---|
| SLIDING SHEETS: | | | YES      NO | | YES      NO | | YES      NO | |
| ERGONOMIC BELTS: | | | YES      NO | | YES      NO | | YES      NO | |
| SLIDING BOARDS OR ROLLER BOARDS: | | | YES      NO | | YES      NO | | YES      NO | |
| OTHER: | | | YES      NO | | YES      NO | | YES      NO | |

* **N.B.:** Attach floor plan to assess available space for more equipment and if there is an equipment storage room

**DESCRIPTION OF ROUTINELY USED STRETCHERS/WHEELCHAIRS**

| CHARACTERISTICS AND INADEQUACY SCORE FOR STRETCHERS | SCORE | STRETCHERS | | | | | |
|---|---|---|---|---|---|---|---|
| | | A | B | C | D | E | Total Nr stretchers |
| | | NR | NR | NR | NR | NR | \|_\|_\| |
| Malfunctioning brakes | 1 | | | | | | |
| Not height-adjustable | 2 | | | | | | |
| Side flaps | 2 | | | | | | |
| Needs to be partially lifted | 1 | | | | | | Total score |
| **Column score (Nr of stretchers × sum of scores)** | | | | | | | |

Mean score (**MSSTR**) |_____| = $\dfrac{\text{Total score stretchers}}{\text{Total Nr stretchers}}$

**FIGURE 8.1** *(continued)*    Services sheet. *(continued)*

FONDAZIONE IRCCS CA' GRANDA
Ospedale Maggiore Policlinico
CLINICA DEL LAVORO · MILAN (ITALY)
ERGONOMICS SECTION

| WHEELCHAIR FEATURES AND INADEQUACY SCORE | Score | TYPES OF WHEELCHAIRS | | | | | | | Total Nr wheelchairs \|__\| |
|---|---|---|---|---|---|---|---|---|---|
| | | A Nr | B Nr | C Nr | D Nr | E Nr | F Nr | G Nr | |
| Malfunctioning brakes | 1 | | | | | | | | |
| Non-removable armrests | 1 | | | | | | | | Total wheelchairs score: |
| Cumbersome backrest | 1 | | | | | | | | |
| Width exceeding 70 cm | 1 | | | | | | | | |
| Column score (Nr of wheelchairs × sum of scores) | | | | | | | | | |

Mean score (**MSWh**) = Total wheelchairs score / total Nr wheelchairs \|_____\| **MSWh**

> **MEAN WHEELCHAIRS AND STRETCHERS INADEQUACY SCORE (MSSTR+ MSWh) = \|_____\|**

### DESCRIPTION OF OUTPATIENT ENVIRONMENT/FURNISHINGS WITH EXAM ROOMS

| CHARACTERISTICS AND INADEQUACY SCORE FOR ENVIRONMENT/FURNISHINGS | Score | Environment/furnishings - Rooms | | | | | Total Nr rooms \|__\| |
|---|---|---|---|---|---|---|---|
| | | Room A | Room B | Room C | Room D | Room E | |
| Free space inadequate for use of aids | 2 | | | | | | |
| Exam bed not height-adjustable | 2 | | | | | | |
| Side flaps-exam bed | 1 | | | | | | |
| Part of exam bed needs to be raised manually | 1 | | | | | | |
| Patient armchair height less than 50 cm | 0,5 | | | | | | Total Score rooms |
| Door width < 85 cm | 1 | | | | | | |
| Column score (sum of scores) | | | | | | | |

### DESCRIPTION OF DAY HOSPITAL ENVIRONMENT/FURNISHINGS: TYPES OF ROOMS (DH)

| CHARACTERISTICS AND INADEQUACY SCORE FOR WARD ROOMS | Score | Nr rooms with Nr__ beds | Nr rooms with Nr__ beds | Nr rooms with Nr__ beds | Nr rooms with Nr__ beds | Nr rooms with Nr__ beds | Total Nr DH rooms \|__\| |
|---|---|---|---|---|---|---|---|
| Space between beds or between bed and wall less than 90 cm | 2 | | | | | | |
| Space between foot of bed and wall less than 120 cm | 2 | | | | | | |
| Unsuitable bed that needs to be partially lifted | 1 | | | | | | |
| Space between bed and floor less than 15 cm | 2 | cm | cm | cm | cm | cm | |
| Patient armchair height less than 50 cm | 0,5 | | | | | | Total score DH rooms: |
| Column score (Nr of rooms × sum of scores) | | | | | | | |

Mean environment inadequacy score (**MSENV**) = $\dfrac{\text{total room score} + \text{total DH room score}}{\text{total Nr of rooms} + \text{total Nr of DH rooms}}$

### NOTES

**FIGURE 8.1** *(continued)*  Services sheet.

FONDAZIONE IRCCS CA' GRANDA
OSPEDALE MAGGIORE POLICLINICO
- CLINICA DEL LAVORO – MILAN (ITALY)
ERGONOMICS SECTION

## OUTPATIENT/DAY HOSPITAL SERVICES: HOW TO ASSIGN "VALUES" TO RISK FACTORS

(OP) indicates the sum total patient handling operators present over 24 hours.
(if some operators are unauthorized or exempted from lifting activities and are not assigned to MPH tasks, do not count them as OPs).

Frequency of patient handling tasks (F) indicates the ratio of patient handling maneuvers/tasks sum (both manual and aided) to the number of operators present over 24 hours (OP).

### TOTAL LIFTING DEVICES FACTOR
THE ADEQUACY OF LIFTING DEVICES IS DETERMINED BASED ON THE PERCENTAGE OF TOTAL AIDED LIFTS (ATLPER)
if ATLPER is below 50% = INADEQUACY HIGH
if ATLPER is below 90% but above (or equal to) 50% = MEDIUM INADEQUACY
if ATLPER is at least 90% = INADEQUACY NEGLIGIBLE

### PARTIAL LIFTING DEVICES FACTOR
THE ADEQUACY OF LIFTING DEVICES IS DETERMINED BASED ON THE PERCENTAGE OF PARTIAL AIDED LIFTS (APLPER)
if APLPER is below 90% = INADEQUACY HIGH
if APLPER is at least 90% = INADEQUACY NEGLIGIBLE

### WHEELCHAIRS/STRETCHERS FACTOR
To define the wheelchairs factor, it is necessary to assess the "Mean wheelchairs and stretchers inadequacy score" reported in the data collection sheet (MSSTR + MSWh) as shown in the following table:

| STRETCHERS-WHEELCHAIRS FACTOR | | | |
|---|---|---|---|
| MSSTR + MSWh | 0.0 – 3.33 | 3.34 – 6.66 | 6.67 - 10 |
| ERGONOMIC INADEQUACY | NEGLIGIBLE | MEDIUM | HIGH |

### ENVIRONMENT FACTOR
To assign the environment factor, the "Mean environment score" (MSENV) calculated on the data collection sheet is divided into three equidistant ranges (i.e. low, medium and high inadequacy), as shown in the following table:

| MSENV | 0-2.5 | 2.51-5 | 5.01-7.5 |
|---|---|---|---|
| ERGONOMIC INADEQUACY | NEGLIGIBLE | MEDIUM | HIGH |

### TRAINING FACTOR

| Characteristics observed | ERGONOMIC INADEQUACY |
|---|---|
| Adequate training course held no more than 2 years before the risk assessment, attended by 75% of surgical unit staff | NEGLIGIBLE |
| Training course held more than 2 years before the risk assessment, attended by 75% of surgical unit staff, and tested for effectiveness | NEGLIGIBLE |
| Adequate training course held no more than 2 years before the risk assessment, attended by between 50% and 75% of surgical unit staff | MEDIUM |
| Only explanations (or specific brochures) given to 90% of surgical unit staff, followed by effectiveness testing | MEDIUM |
| TRAINING NOT DELIVERED OR NOT COMPLIANT WITH AFOREMENTIONED CONDITIONS | HIGH |

**FIGURE 8.2** Services sheet legend.

## SUMMARY OF OUTPATIENT/DAY HOSPITAL SERVICES    Date _____

| Hospital | Service | Code |
|---|---|---|

Nr visits/day _____    Average number/day of visits by disabled patients (D) _____  OPERATORS (Op)   |_|

| FREQUENCY OF PATIENT HANDLING TASKS | SUM TL + SUM PL / |_|_|Op | |_|_|,|_| F |
|---|---|---|

Nr totally non-cooperative patients **NC** _____    Nr partially cooperative patients **PC** _____

### "TOTAL LIFTING DEVICES" FACTOR

| PERCENTAGE OF TOTAL AIDED LIFTS |_|_| ATLPER | ERGONOMIC INADEQUACY |
|---|---|
| EQUIPMENT SELDOM USED (ATLPER below 50%) | HIGH |
| EQUIPMENT SOMETIMES USED (ATLPER below 90% but above (or equal to) 50%) | MEDIUM |
| EQUIPMENT ADEQUATELY USED (ATLPER ≥ 90%) | NEGLIGIBLE |

### "PARTIAL LIFTING DEVICES" FACTOR

| PERCENTAGE OF AIDED PARTIAL LIFTS |_|_| APLPER | ERGONOMIC INADEQUACY |
|---|---|
| EQUIPMENT SELDOM USED (APLPER below 90%) | HIGH |
| EQUIPMENT ADEQUATELY USED (APLPER ≥ 90%) | NEGLIGIBLE |

| EXPOSURE LEVEL | TICK EXPOSURE LEVEL | FREQUENCY OF PATIENT HANDLING TASKS (F) |
|---|---|---|
| ABSENT | D = 0 | |
| NEGLIGIBLE | ATLPER ≥ 90%AND APLPER ≥ 90% | |_|_|,|_| F |
| MEDIUM | ATLPER ≥ 90% AND APLPER < 90% | |_|_|,|_| F |
| HIGH | ATLPER < 90% | |_|_|,|_| F |

### OTHER IMPORTANT ASPECTS WITH REGARD TO RISK REDUCTION

| STRETCHERS-WHEELCHAIRS FACTOR | | | | |
|---|---|---|---|---|
| Mean wheelchairs and stretchers inadequacy score (MSSTR+MSWH) | 0.0 – 3.33 | 3.34-6.66 | 6.67-10 | (MSSTR+MSWh): |_|_|,|_| |
| ERGONOMIC INADEQUACY | NEGLIGIBLE | MEDIUM | HIGH | |
| ENVIRONMENT FACTOR | | | | |
| Mean environment score (MSENV) | 0-2.5 | 2.51-5 | 5.01- 7.5 | (MSENV): |_|_|,|_| |
| ERGONOMIC INADEQUACY | NEGLIGIBLE | MEDIUM | HIGH | |
| TRAINING FACTOR | | | | |
| type of training | adequate | Partly adequate | Completely inadequate | |
| ERGONOMIC INADEQUACY | NEGLIGIBLE | MEDIUM | HIGH | |

**FIGURE 8.3**   Summary service.

# 9 Emergency Department
## *Setting Up Preventive Strategies for Manual Patient Handling*

## 9.1 PRESENTATION

As far as manual patient-handling tasks are concerned, emergency services are a major challenge for a variety of reasons. The greatest difficulty lies in putting together an overview of a facility that needs to deliver healthcare services under emergency conditions, with extremely fast reaction times on the part of staff. Another challenge is posed by the extremely variable work scenarios that can be found in emergency departments, in terms of both the wide range of different people and pathologies passing through and the way the structure is organized to cope with so many diverse emergencies and disabilities. All this reflects on the methods and frequency of manual patient-handling activities.

The proposal set forth here aims solely to assess manual patient-handling risk among workers on hand in the Emergency Department under examination, therefore excluding emergency care workers employing ambulances or helicopters. Whereas the literature may help to identify the prevalence or impact of backache on ambulance and helicopter transport staff, no such information is available with respect to Emergency Department staff in hospitals.

The variables encountered in health emergencies and the types of patients turning to emergency services make it unlikely for manual patient-handling risk exposure to be low, since the following circumstances are typically found on a daily basis:

- The need for speed of action (due to concomitant emergencies)
- Noncooperative or disabled patients needing to be transferred horizontally over a considerable distance (from the operator's body), thus generating high biomechanical overload situations
- Patient self-sufficiency/disability often unknown

This is why emergency services are classified as *high risk* for biomechanical overload so that our principal aim is to define a specific preventive plan.

## 9.2  INFORMATION GATHERING: THE DATA COLLECTION SHEET

The approach to describing the organization and environments of the facility is the same as for the other hospital wards and sectors. The data gathering model in fact includes all the factors that can be used to put in place a targeted and detailed preventive strategy—that is,

- Number of visits to the Emergency Department and number of patients needing to be handled (totally or partially) by Emergency Department staff
- Types of maneuvers routinely performed
- Total number of operators assigned to patient-handling activities over 24 hours
- Structural aspects of the work environment
- Patient-handling aids and devices
- Operator training

The most obvious difference between the Emergency Department and other wards is that it is impossible to quantify accurately the number of patients who need to be totally lifted (noncooperative or NC) or partially lifted (partially cooperative or PC) . Therefore, it is necessary to calculate the percentage of patients needing total or partial lifting with respect to the total number of daily visits.

The analysis also examines the specific patient-handling equipment on hand (i.e., height-adjustable stretchers with sliding boards or sheets, minor aids, and wheelchairs). Ergonomically unacceptable situations that could determine more frequent patient-handling activities or biomechanical overload of the spine are assessed.

The last two sections of the data sheet examine the characteristics of the space and furnishings of all the rooms in the Emergency Department and the training of personnel for the specific risk.

On the whole, as stated before, the data collection sheet includes certain aspects and comments designed to gather all the information required to plan remedial actions in the facility. This version is the result of the ongoing cooperation and suggestions of all the operators participating in the multicenter trials conducted by our research unit (Chapter 2 of this book).

### 9.2.1  ORGANIZATIONAL ASPECTS

A detailed analysis of the data collection sheet (Figure 9.1) for detecting risk in the Emergency Department shows that the original approach remains unchanged and calls for information to be gathered via both an interview with the head of the unit and an on-site inspection.

During the interview, after explaining the reason why the data are being collected, the head of the facility (or an "expert" co-worker) is asked to describe the following organizational aspects:

- Operators engaged in patient-handling activities (Op over 24 hours)
- Percentage of incoming patients requiring total or partial lifting by operators

- Type of routine patient-handling tasks
- Staff training (identical to ward staff)

In emergency departments, the principal factor for defining specific patient-handling risk is represented by the average number of daily visits to the facility. This information is easy to find because hospitals produce an official annual report on visits to emergency services.

Next, the investigation goes into the amount of manual patient handling required of operators in the emergency service. The workload is expressed as a percentage of patients needing handling with respect to the total number of daily visits.

During the interview with the head of the facility, all the routine patient-handling operations are required (whether manual or aided), as described in the relative tables ("Table A" and "Table B" on the data collection sheet; see Figure 9.1).

The reason why patient-handling operations are described is to define the types and numbers of operations carried out using patient aids or devices (**AMPER**) based on the ratio of aided to total handling tasks, as shown:

Percentage of aided MPH (manual patient-handling) tasks = Table B (aided patient-handling tasks)/Table A (manual + aided handling tasks)

During the on-site inspection, the ergonomic adequacy of the patient-handling equipment and aids, and the environments in which the activities are performed, is assessed.

### 9.2.2 ON-SITE INSPECTION: DESCRIPTION OF PATIENT-HANDLING EQUIPMENT

The patient-handling equipment is assessed based on ergonomic adequacy, number, and actual use; more specifically, the analysis covers examination beds and/or waiting room beds and their ergonomic adequacy (Table 9.1). Next comes a detailed description of any minor aids such as sliding boards or sheets.

"Lifting" equipment is difficult to use in emergency services due to the most frequent type of patient handling involved (i.e., from one flat surface to another, such as from stretcher to bed and vice versa) and to the fact that emergency situations need to be dealt with at the utmost speed.

If the Emergency Department does have a patient hoist, this will be stated in the section titled "Note."

Stretchers and wheelchairs are analyzed using the same approach—namely, by attributing a score to their ergonomic inadequacy features to obtain, at the end, the "mean wheelchairs and stretchers inadequacy score" for stretcher (MSSTR) + mean score for wheelchair (MSWh).

In fact, such equipment is used both as a device for transporting disabled patients (stretchers and wheelchairs) and as a surface for transferring them (stretchers) and must be ergonomically adequate in order to be used as an aid in transfer operations (without operators adopting awkward postures).

**TABLE 9.1**

**Description of Ward Beds and Exam Beds**

| Describe Type of Bed | No. | Electrically Powered | | Mechanical Pedal Operated | | No. Sections | | | Manually Lifted Head or Foot Section | |
|---|---|---|---|---|---|---|---|---|---|---|
| Bed A: | | Yes | No | Yes | No | 1 | 2 | 3 | Yes | No |
| Bed B: | | Yes | No | Yes | No | 1 | 2 | 3 | Yes | No |
| Bed C: | | Yes | No | Yes | No | 1 | 2 | 3 | Yes | No |
| Bed D: | | Yes | No | Yes | No | 1 | 2 | 3 | Yes | No |
| Exam bed A: | | Yes | No | Yes | No | 1 | 2 | 3 | Yes | No |
| Exam bed B: | | Yes | No | Yes | No | 1 | 2 | 3 | Yes | No |
| Exam bed C: | | Yes | No | Yes | No | 1 | 2 | 3 | Yes | No |

### 9.2.3 ON-SITE INSPECTION: ASSESSMENT OF SPACE AND FURNISHINGS

During the on-site inspection, the organizational aspects reported during the interview with the head of the department will be verified, and the exam rooms and waiting rooms will also be analyzed with regard to space and furnishings. In general, Emergency Department bathrooms are seldom used by disabled/noncooperative patients, and consequently the analysis does not include them.

More specifically, nonergonomic conditions and characteristics are listed in Table 9.2, and these contribute to defining the mean environment inadequacy score. Also, the overall "inadequacy" of the environment is expressed by a mean score for environment (MSENV), and calculated as illustrated:

Mean "inadequacy" score for environment (MSENV) = total room score/total number of rooms

**TABLE 9.2**

**Description of Environment**

**Characteristics Analyzed**

Inadequate space for using patient-handling aids

Exam bed not height adjustable

Side flaps or rails prevent stretcher from being brought close to bed

Part of exam bed needs to be raised manually

Door width < 85 cm

**TABLE 9.3**
**Ergonomic Inadequacy of Furniture**

| Percentage of Aided MPH Tasks | Ergonomic Inadequacy |
|---|---|
| AMPER ≤ 50% | High |
| 50% > AMPER < 90% | Medium |
| AMPER ≥ 90% | Negligible |

## 9.3 RISK EXPOSURE ASSESSMENT: FINAL OVERVIEW

As stated previously, since activities are performed under "urgent/emergency" conditions, **Emergency Departments should be regarded as being at risk for manual patient-handling disorders.**

The description of each individual risk factor must be used to reduce inadequacy levels and acquire data for planning environmental improvements/remedial action.

The information reported on the data collection sheet is therefore vital for evaluating the adequacy of the patient-handling aids present and the specific organization of the activities carried out in this area, expressed as the percentage of aided patient lifts (AMPER).

In accordance with the rationale used in previous chapters, this percentage breaks down "ergonomic inadequacy" into three levels (Table 9.3):

All the information reported on the data collection sheet is then quantified in the final overview, which will indicate

- Total number of visits
- Number of operators assigned daily to patient-handling duties
- Number of NC patients/day
- Adequacy of patient-handling aids (three levels)
- Adequacy of patient-transfer equipment: wheelchairs and stretchers (three levels)
- Adequacy of environment (three levels)
- Adequacy of staff training (three levels)

This makes it easier to define remedial actions; based on a known workload defined by the number of patients to be moved and the number of operators per day, this identifies and emphasizes the most significant risk factors (Table 9.4).

**TABLE 9.4**
**High, Medium, Low Ergonomic Inadequacy Criteria for Different Risk Factors in the Emergency Department**

| Risk Factor | Level of Ergonomic Inadequacy |
|---|---|
| Equipment factor | High (for AMPER $\leq$ 90%) |
| | Medium (for 50% $\geq$ AMPER $\leq$ 90%) |
| | Low (for AMPER $\geq$ 90%) |
| Stretchers–wheelchairs factor | High (for MSSTR + MSWh = 6.67 – 10) |
| | Medium (for MSSTR + MSWh = 3.34 – 6.66) |
| | Low (for MSSTR + MSWh = 0 – 3.33) |
| Environment factor | High (for MSENV = 4.01 – 6) |
| | Medium (for MSENV = 2.01 – 4) |
| | Low (for MSENV = 0 – 2) |
| Training factor | High (training not delivered or not compliant with conditions indicated below) |
| | Medium (training course held no more than 2 years before risk assessment, attended by between 50% and 75% of department staff) |
| | Medium (only training or distribution of specific brochures to 90% of department staff, followed by effectiveness testing) |
| | Low (adequate training course held no more than 2 years before risk assessment, attended by 75% of department staff) |
| | Low (training course held more than 2 years before risk assessment, attended by 75% of department staff, and tested for effectiveness) |

**COMPLETE FORM: RISK ASSESSMENT FOR MANUAL PATIENT HANDLING
IN EMERGENCY DEPARTMENT**

HOSPITAL: _____ Date _____

EMERGENCY DEPARTMENT: _____ Code _____

### 1. INTERVIEW

| Nr OF OPERATORS ENGAGED IN MPH: indicate the total number of operators per job category | | | |
|---|---|---|---|
| nurses: | orderlies (ASA/OTA/OSA): | nurses aides: | Other: |

**Nr OF OPERATORS PERFORMING MPH TASKS OVER 3 SHIFTS:** indicate number of operators on duty per shift:

| SHIFT | morning | afternoon | night |
|---|---|---|---|
| Shift schedule: (00:00 to 00:00) | from_____to_____ | from_____to_____ | from_____to_____ |
| Nr of operators over entire shift | | | |
| (A) Total operators over entire shift | | | |

**Nr of PART-TIME OPERATORS:** indicate the exact number of hours worked and calculate them as unit fractions (in relation to the overall duration of the shift)

| Nr of part-time operators present | Hours worked in shift: (from 00:00 to 00:00) | Unit fraction | (Unit fraction by Nr of operators) |
|---|---|---|---|
| | from_____to_____ | | |
| (B) Total operators (as unit fractions) present by shift duration = | | | |

| TOTAL Nr OF OPERATORS ENGAGED IN MPH OVER 24 HOURS (Op): add the total number of operators present over the entire shift (A) to the total number of part-time operators (B) | | Op |
|---|---|---|

Do the nurses in the Emergency Department also work as transport nurses? NO YES Emergency calls _____

**AVERAGE NUMBER OF VISITS/DAY:** _____

Indicate the percentage of patients visiting Emergency Department who need full or partial lifting by staff members, versus the total number of visits/day  |__|__| %

| TABLE A: Describe routine patient handling operations (manual and aided) | | | |
|---|---|---|---|
| wheelchair/exam bed (or stretcher) | |__| | exam bed (or stretcher)/wheelchair | |__| |
| stretcher/bed | |__| | exam bed/stretcher | |__| |
| stretcher/RX table or similar | |__| | RX table or similar/stretcher | |__| |
| wheelchair/RX table or similar | |__| | RX table or similar/wheelchair | |__| |
| stretcher/bed | |__| | Other _____ | |__| |
| Other _____ | |__| | Other _____ | |__| |
| TOTAL A (total **types** of routine patient handling tasks): | | | |

| OPERATOR TRAINING | | | | | |
|---|---|---|---|---|---|
| **TRAINING** | | | **INFORMATION** | | |
| Attended theoretical/practical course | ☐ YES | ☐ NO | Training only on how to use equipment | ☐ YES | ☐ NO |
| if YES, how many months ago? and how many hours/operator | Months _____ hours _____ | | Only provided brochures on MPH | ☐ YES | ☐ NO |
| if YES, how many operators? | | | if YES, how many operators? | | |
| Was EFFECTIVENESS measured and documented in writing? | | ☐ YES | | ☐ NO | |

**FIGURE 9.1** Emergency service sheet. *(continued)*

### 2. ON-SITE INSPECTION

**EQUIPMENT FOR DISABLED PATIENT LIFTING/TRANSFERS**

| Height-adjustable examination beds Nr= \|__\|__\| | | Emergency department waiting room: Height-adjustable examination beds Nr= \|__\|__\| | | | |
|---|---|---|---|---|---|
| DESCRIBE TYPE OF BED | Nr | Electric adjustable | Mechanical adjustable | Nr of sections | Manual lifting of bed head or foot |
| BED A: | | YES  NO | YES  NO | 1  2  3  4 | YES  NO |
| BED B: | | YES  NO | YES  NO | 1  2  3  4 | YES  NO |
| BED C: | | YES  NO | YES  NO | 1  2  3  4 | YES  NO |
| BED D: | | YES  NO | YES  NO | 1  2  3  4 | YES  NO |
| EXAM BED A: | | YES  NO | YES  NO | 1    2 | YES  NO |
| EXAM BED B: | | YES  NO | YES  NO | 1    2 | YES  NO |
| EXAM BED C: | | YES  NO | YES  NO | 1    2 | YES  NO |

| EQUIPMENT DESCRIPTION | | Nr | Lack of essential requirements | Lack of adapt-ability to patients or environment | Lack of maintenance |
|---|---|---|---|---|---|
| SLIDING SHEETS | | | YES  NO | YES  NO | YES  NO |
| ERGONOMIC BELTS: | | | YES  NO | YES  NO | YES  NO |
| SLIDING BOARDS: | | | YES  NO | YES  NO | YES  NO |
| OTHER: | | | YES  NO | YES  NO | YES  NO |

\* N.B.: Attach floor plan to assess available space for more equipment and if there is an equipment storage room

| **TABLE B:** Indicate patient handling operations using aids | | | |
|---|---|---|---|
| bed (or stretcher)/wheelchair | \|__\| | bed (or stretcher)/wheelchair | \|__\| |
| stretcher/exam bed | \|__\| | bed/stretcher | \|__\| |
| stretcher/RX table or similar | \|__\| | RX table or similar/stretcher | \|__\| |
| wheelchair/RX table or similar | \|__\| | RX table or similar/wheelchair | \|__\| |
| stretcher/ward bed | \|__\| | Other _____ | \|__\| |
| Other _____ | \|__\| | Other _____ | \|__\| |
| TOTAL B (total types of routine aided patient handling tasks): | | | |

**PERCENTAGE OF AIDED MPH TASKS (TABLE B/TABLE A) = \|__\|__\| AMPER**

**DESCRIPTION OF ROUTINELY USED STRETCHERS**

| CHARACTERISTICS AND INADEQUACY SCORE FOR STRETCHERS | Score | TYPE OF STRETCHERS | | | | | Total Nr stretchers |
|---|---|---|---|---|---|---|---|
| | | A | B | C | D | E | \|__\|__\| |
| | | Nr | Nr | Nr | Nr | Nr | |
| Malfunctioning brakes | 1 | | | | | | |
| Not height-adjustable | 2 | | | | | | |
| Side flaps | 2 | | | | | | Total stretchers score |
| Needs to be partially lifted | 1 | | | | | | |
| Column score (Nr stretchers × sum of scores) | | | | | | | |

Mean score **MSSTR** \|__\|. \|__\|__\| = Total score stretchers / Total Nr stretchers

**FIGURE 9.1** *(continued)*  Emergency service sheet. *(continued)*

FONDAZIONE IRCCS CA' GRANDA
OSPEDALE MAGGIORE POLICLINICO
CLINICA DEL LAVORO – MILAN-ITALY
ERGONOMICS SECTION

**DESCRIPTION OF ROUTINELY USED WHEELCHAIRS**

| WHEELCHAIRS: | Score | TYPE OF WHEELCHAIRS | | | | | | | |
|---|---|---|---|---|---|---|---|---|---|
| CHARACTERISTICS AND INADEQUACY SCORE FOR STRETCHERS | | A Nr | B Nr | C Nr | D Nr | E Nr | F Nr | G Nr | Total Nr wheelchairs \|__\| |
| Malfunctioning brakes | 1 | | | | | | | | |
| Non-removable armrests | 1 | | | | | | | | |
| Cumbersome backrest | 1 | | | | | | | | Total score Wheelchairs: |
| Width exceeding 70 cm | 1 | | | | | | | | |
| **Column score** (Nr wheelchairs × sum of scores) | | | | | | | | | |

Mean score (**MSWh**) = Total wheelchairs score / total Nr wheelchairs \|_____\| **MSWh**

---

**MEAN WHEELCHAIRS AND STRETCHERS INADEQUACY SCORE (MSSTR+ MSWh) = \|_____\|**

---

**DESCRIPTION OF EXAM ROOMS ENVIRONMENT/FURNISHINGS**

| CHARACTERISTICS AND INADEQUACY SCORE FOR ENVIRONMENT/FURNISHINGS | | Environment/furnishings – Exam Rooms | | | | | |
|---|---|---|---|---|---|---|---|
| | | Room A | Room B | Room C | Room D | Room E | |
| Free space inadequate for use of aids | 2 | | | | | | Total Nr rooms \|_\|_\| |
| Exam bed/stretcher not height-adjustable | 1 | | | | | | |
| Side flaps-exam bed/stretcher | 1 | | | | | | |
| Part of exam bed needs to be raised manually | 1 | | | | | | |
| Door width <85 cm | 1 | | | | | | Score Total rooms |
| **Column score** (sum of scores) | | | | | | | |

Mean environment inadequacy score (**MSENV**) = $\dfrac{\text{Total room score}}{\text{Total Nr rooms}}$

**NOTES:**

_____
_____
_____
_____

**FIGURE 9.1** *(continued)* Emergency service sheet.

# 10 Procedures and Tools for Choosing Adequate Equipment

## 10.1 INTRODUCTION

The literature regarding strategies for the prevention of manual patient-handling risk agrees that adopting *adequate* equipment and aids is one of the mainstays of this highly complex area. The term "aid" is typically used to describe devices or equipment used to reduce biomechanical overload of the spine among healthcare workers performing total patient-lifting operations, such as transferring the patient from a bed to a wheelchair, or during partial patient-shifting operations, such as repositioning a cooperative patient in bed. Therefore, examples of equipment and aid include wheelchairs, hospital beds, stretchers, hoists, bathing aids, and "minor aids."

The choice and consequent use of equipment is based on a process that combines identifying the most suitable type of aid (of which the relevant requirements will be explored later) with providing practical training/education for the personnel who use the equipment.

However, before actually deploying equipment, it is essential to consider (Nelson et al. 2006) the possible existence of numerous barriers to use—for example,

- Patient aversion to lifting devices (especially hoists)
- Equipment unstable or difficult to use
- Inadequate equipment storage space
- Poor equipment maintenance
- Excessively slow utilization time
- Too few aids
- Lack of training (especially in situations with high staff turnover)
- Inadequate turning/maneuvering space
- Excessively high costs

These considerations also stem from the fact that all too often, equipment is purchased that fails to meet the needs of the specific ward/facility, and that sometimes adequate equipment is present but nonetheless is not used by workers. In the authors' experience, one of the reasons for inadequate aids is poor coordination between hospital staff members, especially with procurement departments. The main priority of procurement offices is to achieve cost savings, which does not always go hand in hand with the need to purchase adequate patient-handling aids.

**TABLE 10.1**
**Ergonomic Equipment Requirements**

| Preliminary Requirements | Specific Requirements |
| --- | --- |
| Safe (for operator/patient) | Adaptable to function |
| Comfortable for patient | Adaptable to environment |
| Easy to use | Adaptable to patient |
| Low applied physical effort | |

ISO TR 12296 (2012) has stressed the importance of defining a procedure for choosing adequate aids, based on an analysis of

- The type of patients routinely handled
- The organizational aspects of shift work
- The space and furnishings in the area where aids are to be used
- Priorities, based on the degree of biomechanical overload caused by specific patient-handling tasks
- The individual ergonomic characteristics that the equipment for a specific unit/ward must have, as illustrated in Table 10.1

In order to ensure an effective strategy for reducing manual patient-handling risk, the following steps must be taken:

Step 1. *Risk analysis* to identify:
  - Priorities
  - Detailed analysis of types of patient-handling activities to be aided
  - Detailed analysis of types of patient disabilities
  - Detailed analysis of environments in which aids are to be utilized
Step 2. Preparation of a *list of ergonomic requirements* for equipment purchases, which should include:
  - Preliminary ergonomic equipment requirements
  - Specific ergonomic equipment requirements
  - Request to test the equipment in the specific ward/facility for at least 2 weeks, to provide both *technical training* on the use of the equipment (by the manufacturers) and *training in the correct use of the equipment for risk reduction purposes* (by professionals on the hospital staff responsible for risk reduction training, i.e., ergocoaches, back-care advisers, etc.).
Step 3. Once the equipment has been purchased by the ward/facility, a procedure for checking that it is being used needs to be implemented. In order to help choose the right equipment, a number of specific protocols have been put together for wards (Figure 10.1), surgical units (Figure 10.2), and outpatient services (Figure 10.3). Each protocol begins with a summary of the main factors to be taken into consideration (all of which can be found on the MAPO risk detection form):

- *Summary* of the scores for all the risk factors contributing to define the risk level in the specific ward
- *Types of patients* needing to be moved—that is, either totally noncooperative (with no motor abilities) or partially cooperative (with residual motor abilities)
- Types of patient lifting or repositioning operations requiring equipment/aids
- Percentage of total lifting or partial repositioning still needing to be aided

The second half of the protocol examines the different types of equipment/aids for the specific areas of the hospital (e.g., wards, surgical units, outpatient services). Each form also includes a second step, to be followed after delivery of the test equipment, which serves to assess the ergonomic requirements of the equipment.

As recommended by several guidelines for managing manual patient-handling risk (RCN 2003a; Nelson 2009), each equipment requirement is evaluated by means of a visual-analog scale of 1 to 5, where 1 indicates that the specific requirement is almost absent, and 5 indicates that the ergonomic requirement is fully present.

The final part of the questionnaire recalculates the percentage of total lifts or partial aided lifts and thus recalculates the MAPO (*movimentazione e assistenza pazienti ospedalizzati*—movement and assistance for hospitalized patients) exposure level.

These forms serve several purposes: besides guiding the choice of the most adequate aid for the ward/unit under examination, they enable a prediction to be made of the expected impacts on risk reduction; those responsible for allocating financial resources will thus be able to base their decisions on objective criteria.

## 10.2 APPROACHES TO CHOOSING ADEQUATE LIFTING AIDS

"Ergonomic" lifting aids are electrically powered devices that can completely lift the patient. Table 10.2 lists the main features of a patient-lifting hoist that should be present to ensure that the preliminary requirements of safety, comfort, ease of use, and low physical effort applied are met. These preliminary requirements, in addition to the various features of the hoist, will determine the choice of the most suitable device for the ward or facility in which it is to be used. It is worth pointing out that the equipment should also comply with the technical and ergonomic requirements laid down by ISO 10535 (2006).

Once the preliminary requirements have been assessed, it is necessary to check whether the equipment is specifically *adaptable* to the *patient/type of handling/ environment,* based on a more detailed organizational analysis. Hoists may be on wheels or ceiling mounted. With regard to the physical effort required to pull–push wheeled hoists, the literature (Marras 2008; Hignett and Keen 2005) now places much emphasis on minimizing biomechanical overload while using ceiling-mounted lifting devices. Moreover, the commonly limited space in which wheeled hoists are pushed and pulled may lead to specific overload when changing direction.

**TABLE 10.2**

**Preliminary Requirements for Hoist and Main Aspects to be Observed**

| Preliminary Requirements | Hoist Aspects | | |
|---|---|---|---|
| Safe for operator | • Emergency lowering mechanism | Yes | No |
| | • Braking system responsive and readily accessible | Yes | No |
| | • Adjustments smooth | Yes | No |
| | • Dual controls (for patient and caregiver) | Yes | No |
| Safe for patient | • Automatic stop for overload or obstacles during lowering | Yes | No |
| | • Security of sling attachment system | Yes | No |
| | • Maximum lifting weight clearly indicated | Yes | No |
| | • Controls clearly identified | Yes | No |
| Comfortable for patient | • Sling adjustable to different patient sizes | Yes | No |
| | • Patient in lifted position without excessive compression | Yes | No |
| | • Patient in lifted position without swinging | Yes | No |
| Low physical effort applied | • Wheels with low friction | Yes | No |
| | • Low weight of structure, including frame | Yes | No |
| | • Electric lifting mechanism | Yes | No |
| | • Correct patient position during lifting (with sling) | Yes | No |
| | • Adequate shape and height of pulling/pushing handles | Yes | No |
| Easy to use | • Controls (clearly identified—mode of operation) | Yes | No |
| | • Type of sling with easy method of positioning removal | Yes | No |
| | • Designed to be used by a single operator | Yes | No |

This applies to long-term care wards for noncooperative/disabled patients and where there are no short-term plans for reassigning the rooms to a different use. Lifting devices suspended from the ceiling are particularly recommended for extended care facilities, spinal units, or neurorehabilitation gyms, as well as in intensive care and radiology departments.

Hoists are used for helping patients to be moved

• From bed to wheelchair and vice versa
• From wheelchair to toilet and vice versa
• From wheelchair to bath and vice versa
• From bed to stretcher and vice versa
• To and from the floor

Lifting devices can also help to transport patients over short distances (e.g., from bedroom to bathroom). Ceiling-mounted lifting aids require structural work (on ceiling and doors) as well as an adequate load-bearing structure (ceiling).

If taken into consideration in the planning of a new facility, such devices generate a positive cost:income ratio; they are used because they are always available, do not occupy space around the patient's bed, or hamper diagnostic equipment, which sometimes calls for patients to be placed in awkward positions, such as for MRIs in radiology departments.

## 10.3 APPROACHES TO CHOOSING HEIGHT-ADJUSTABLE STRETCHERS

Height-adjustable stretchers have satisfactorily met the needs of hospitals as they have evolved over the past 15 years, during which average stays have dropped to only a few days during acute stages or for diagnostic purposes, and where patients are most commonly transported to other units for tests or surgical procedures, thus calling for transfers from bed to stretcher and from stretcher to examination bed.

To move patients from one surface to another, the stretcher must not only be height-adjustable, but also equipped with sliding sheets or boards. Table 10.3 lists the main features that stretchers should have to ensure that the preliminary ergonomic requirements of safety, comfort, ease of use, and low physical effort applied are met.

**TABLE 10.3**
**Preliminary Requirements for Adjustable Stretchers and Main Aspects to be Observed**

| Preliminary Requirements | Main Aspects of Stretcher | | |
|---|---|---|---|
| Operator safety | • Efficient braking system for all four wheels | Yes | No |
| Patient safety | • Efficient braking system for all four wheels | Yes | No |
| | • Automatic overload stop device | Yes | No |
| | • Side rails can be completely lowered | Yes | No |
| Patient comfort | • Controls determine smooth movements | Yes | No |
| | • Adequate for all patient sizes | Yes | No |
| | • Two-section design | Yes | No |
| Low physical effort applied | • Electric controls | Yes | No |
| | • Low-friction/pivoting wheels | Yes | No |
| | • Low weight of frame | Yes | No |
| | • Five wheels | Yes | No |
| | • No sections requiring manual lifting | Yes | No |
| Easy to use | • Controls clearly identified | Yes | No |
| | • No side flaps/rails | Yes | No |
| | • Space under stretcher to allow hoist to be used | Yes | No |

## 10.4   APPROACHES TO CHOOSING PATIENT BEDS

Most patient care activities are carried out at the bedside. It is therefore crucial for the bed to meet certain basic requirements. Electrically powered, height-adjustable beds reduce the risk of adopting awkward postures, reduce mechanical stress during certain transfers, and allow elderly or partially cooperative patients to adjust the bed (while in a nonergonomic bed, the patient needs someone to lift/help him or her).

Beds with three electrically powered sections reduce operator risk, improve the quality of care, and avoid patients needing to be repositioned frequently or pulled up in bed. Beds with sides or rails that can be completely lowered are essential to avoid hampering operators during bedside care tasks. Sides or rails divided into two sections also give partially independent patients something to lean on when moving from sitting to standing (and vice versa), thus reducing the number of times they need to be assisted and enhancing their independence. Side rails with horizontal bars are inadequate: They are unsafe and uncomfortable for patients because of the risk of catching or trapping a paralyzed limb.

Adjustment mechanisms must leave enough room (at least 15 cm) under the bed so as to position a hoist. Moreover, if patient beds need pushing or pulling, it is essential to acquire types with a lightweight frame and a fifth wheel at the center of the bed to facilitate turning corners. It is also worth pointing out that the equipment should comply with the technical and ergonomic requirements laid down by EN IEC 60601-2-52.

As in the case of other equipment, Table 10.4 lists the main aspects of the bed with respect to meeting preliminary ergonomic requirements.

## 10.5   APPROACHES TO CHOOSING MINOR AIDS

"Minor aids" are devices used primarily for partially handling noncooperative patients. Minor aids include

- **Ergonomic belts** have handles attached to the sides and back; the belt is placed around the waist of partially cooperative patients and enables the operator to guide the patient's movements and posture. Ergonomic belts are recommended for use in wards housing patients who can use at least one lower limb (e.g., hemiplegic, elderly and somewhat frail, hip fractures, or hip replacements). These devices are used to transfer patients from sitting to standing positions and vice versa. However, ergonomic belts should not be used to *lift* patients but rather to "steer" the patient's weight toward the healthy leg (so that the patient is able to make use of any residual motor abilities). The specific ergonomic requirements for ergonomic belts are
  - At least two lateral and vertical handgrips
  - Handles with 10-cm grip space for the operator
  - Comfortable for patient (possibly padded)
  - Safe buckle mechanism
  - Adaptable to different patient builds (or available in multiple sizes)

**TABLE 10.4**

**Preliminary Requirements for Adjustable Beds and Main Aspects to be Observed**

| Preliminary Requirements | Main Aspects of Bed | | |
|---|---|---|---|
| Operator safety | • Efficient braking system | Yes | No |
| | • Side/rails (and other mobile parts of bed) not at risk for crushing hands or fingers | Yes | No |
| Patient safety | • Certain adjustments not accessible to patient | Yes | No |
| | • Efficient braking system | Yes | No |
| | • Automatic overload stop device | Yes | No |
| | • Sides/rails not at risk for catching/trapping limbs | Yes | No |
| Patient comfort | • Controls determine smooth movements | Yes | No |
| | • All sections adequate for all patient sizes | Yes | No |
| | • Four-section design | Yes | No |
| | • Height adjustable | Yes | No |
| Low physical effort applied | • Electric controls | Yes | No |
| | • Low-friction/pivoting wheels | Yes | No |
| | • Low weight of frame | Yes | No |
| | • Five wheels | Yes | No |
| | • No sections requiring manual lifting | Yes | No |
| | • Adjustable to Trendelemburg/anti-Trendelemburg position | Yes | No |
| Easy to use | • Controls clearly identified | Yes | No |
| | • No side flaps/rails | Yes | No |
| | • Space under bed to allow hoist to be used | Yes | No |

- **Sliding sheets and sliding boards** are aids consisting of nylon sheets that are placed under the patient to reduce friction between the patient's body and the surface to which the patient needs to be transferred. These aids help to slide the supine patient from one surface to another without lifting (e.g., from bed to stretcher, or from bed to operating table); they are particularly useful for repositioning, turning, and transferring patients at the bedside. The main requirements for such aids are
  - They must be easy to use, especially when placing under the patient; therefore, they should be made out of a very thin, highly gliding fabric.
  - Minimal applied physical force must be required to slide the patient's body over the surface below.
  - The sheets must be disposable or, alternatively, must be able to be replaced and sterilized.

**TABLE 10.5**

**Preliminary Requirements for Wheelchairs and Main Aspects to be Observed**

| Preliminary Requirements | Aspects of Wheelchair |
|---|---|
| Operator and patient safety | • Braking system responsive and readily accessible<br>• All sharp metal edges must be covered with plastic (even when armrests are removed)<br>• The base must be stable to prevent tipping over |
| Patient comfort | • Position with patient sitting<br>• Adequate adjustable footrest position<br>• Separate leg rests |
| Low physical effort applied | • Type of wheel base<br>• Weight of structure with support system<br>• Handles adequate for pushing and pulling |
| Easy to use | • Readily removable armrests (mandatory)<br>• Easy to apply brakes<br>• Easy to adjust position of footrests<br>• Easy to perform other operations (e.g., remove large rear wheels) |

Sliding sheets and boards can also be used in wards and units where patients need to be transferred smoothly from one surface to another (combined with ergonomic stretchers, e.g., in general/surgical/orthopedics wards, etc.).

## 10.6   APPROACHES TO CHOOSING WHEELCHAIRS

Besides meeting standard ergonomic requirements (i.e., removable arm and foot rests and all other patient comfort accessories, appropriate backrest height), wheelchairs must also be chosen to suit not only the space and furnishings of the ward/facility in which they are to be used, but also any other patient-handling aids such as wheeled hoists.

As for the other equipment, Table 10.5 lists the preliminary requirements for wheelchairs. Table 10.6 shows the correlations between certain elements of the wheelchair and the organizational and structural aspects of the ward (obtained from the MAPO risk detection form) that are needed to analyze the extent to which the wheelchair and/or commode is adaptable to the patient, the function, and the environment.

## 10.7   APPROACHES TO CHOOSING PATIENT WASHING AIDS

The choice of patient washing aids must consider the relevant preliminary ergonomic requirements; as well, however, any assessment of "adequacy" requirements must take the following aspects into account:

- *Need for patient washing aids,* depending on the duration of the disabled patient's stay in the ward/facility (such aids are particularly important in geriatric institutions and long-term rehabilitation units)

## TABLE 10.6
## Adaptability Requirements for Wheelchairs and Main Aspects to be Observed

| Wheelchair | Aspects of Wheelchair |
|---|---|
| Type | • Patient disability; time spent in wheelchair (*adaptability to patient*) |
|  | • Type of operation to be aided (*adaptability to function*) |
|  | • Can/cannot be self-propelled by patient (*adaptability to patient*) |
|  | • Distances covered (length/friction) (*adaptability to function/environment*) |
| Base | • Free space in environment (*adaptability to environment*) |
|  | • Width of doors (*adaptability to environment*) |
|  | • Can be combined with other aids (e.g., hoist: maximum depth and width with hoist base) (*adaptability to environment*) |
| Removable armrests | • Types of maneuvers to be aided (*adaptability to function*) |
|  | • Types of patients, e.g., hemiplegic subjects do not always need removable armrests (*adaptability to function*) |

- *Type of patient disability*
- *Presence of adequate space* for the installation of a disabled shower (suitable only for partially cooperative patients) or "disabled bath," or use of "shower stretchers" (suitable for both totally noncooperative and partially cooperative patients)

FONDAZIONE IRCCS CA' GRANDA
OSPEDALE MAGGIORE POLICLINICO
CLINICA DEL LAVORO – MILAN (ITALY)
ERGONOMICS SECTION

## WARDS: DATA COLLECTION SHEET FOR CHOICE OF AIDS

## 1. DATA FROM WARD DATA COLLECTION SHEET

| HOSPITAL: | WARD: | WARD CODE: |
|---|---|---|
| Nr BEDS: | AVERAGE HOSPITAL STAY (days): | DATE: |
| **% OF AIDED TOTAL LIFTING OPERATIONS (%ATL)** | | |
| **% OF AIDED PARTIAL LIFTING OPERATIONS (%APL)** | | |

$$\text{MAPO} = (|_____| \times |\_\_\_\_| + |_____| \times |\_\_\_\_|) \times |\_\_\_\_| \times |\_\_\_\_\_| \times |\_\_\_\_| =$$

INDEX   NC/OP   LF   PC/OP   AF   WF   EF   TF

### MANUAL PATIENT HANDLING TASKS

| MANUAL HANDLING: describe routine tasks involving total or partial patient lifting | Total lifting (TL) WITHOUT EQUIPMENT | | | Partial Lifting (PL) WITHOUT EQUIPMENT | | |
|---|---|---|---|---|---|---|
| indicate the **number of tasks** per shift involving manual patient handling | morning | afternoon | night | morning | afternoon | night |
| | **A** | **B** | **C** | **D** | **E** | **F** |
| ☐ **pulling up in bed** | ☐☐☐☐ | ☐☐☐☐ | ☐☐☐☐ | ☐☐☐☐ | ☐☐☐☐ | ☐☐☐☐ |
| ☐ **turning over in bed (to change position)** | | | | ☐☐☐☐☐ | ☐☐☐☐ | ☐☐☐☐☐ |
| ☐ **bed-to-wheelchair and viceversa** | ☐☐ | ☐☐ | ☐☐ | ☐☐ | ☐☐ | ☐☐ |
| ☐ **lifting from seated to upright position** | | | | ☐☐ | ☐☐ | ☐☐ |
| ☐ **bed-to-stretcher and viceversa** | ☐☐ | ☐☐ | ☐☐ | ☐☐ | ☐☐ | ☐☐ |
| ☐ **wheelchair-to-toilet and viceversa** | ☐☐ | ☐☐ | ☐☐ | ☐☐ | ☐☐ | ☐☐ |
| ☐ **other** | ☐☐ | ☐☐ | ☐☐ | ☐☐ | ☐☐ | ☐☐ |
| ☐ **other** | ☐☐ | ☐☐ | ☐☐ | ☐☐ | ☐☐ | ☐☐ |
| TOTAL: calculate the total for each column | | | | | | |
| **Number of total (TL) or partial (PL) manual lifting tasks** | **A+B+C = TL** | | | **D+E+F=PL** | | |

**FIRST STEP: CHOICE OF AIDS FOR RISK REDUCTION**

| ☐ ADJUSTABLE STRETCHERS (Nr = \|\_\_\|\_\_\| ) | ☐ SLIDING BOARDS (Nr \|\_\_\|\_\_\| ) |
|---|---|
| **REQUIREMENTS OF HOISTS :** | **TYPE AND NUMBER OF SLINGS :** |
| ☐ CEILING MOUNTED HOIST  Nr = \|\_\_\|\_\_\| | _____   _____ |
| ☐ WHEELED HOIST  N°= \|\_\_\|\_\_\| | TYPE 1 Nr = \|\_\_\|TYPE 2 Nr = \|\_\_\|TYPE 2 Nr = \|\_\_\| |
| ☐ STANDING HOIST  N°= \|\_\_\|\_\_\| | |
| ☐ ADJUSTABLE AND ERGONOMIC BEDS (N°= \|\_\_\|\_\_\| ) | |
| ☐ SLIDING SHEETS (N°= \|\_\_\|\_\_\| )   SIZE _____   TYPE _____ | |
| ☐ ERGONOMIC BELTS (N°= \|\_\_\|\_\_\| ) SIZE _____   TYPE _____ | |
| OTHER: _____ _____(N°= \|\_\_\|\_\_\| ) | |
| ☐ BATHING STRETCHER | ☐ HEIGHT ADJUSTABLE BATH |
| ☐ BATHING CHAIR | ☐ SHOWER FLOOR |
| ☐ WHEELCHAIRS          TYPE_____ | (N°= \|\_\_\|\_\_\| ) |

**FIGURE 10.1**   Ward sheet choice aids. *(continued)*

FONDAZIONE IRCCS CA' GRANDA
OSPEDALE MAGGIORE POLICLINICO
CLINICA DEL LAVORO – MILAN | ITALY
ERGONOMICS SECTION

**SECOND STEP: HOW TO ASSESS ERGONOMIC REQUIREMENTS OF AIDS**

**TYPE OF EQUIPMENT:** _____

| PRELIMINARY REQUIREMENTS: VALUTARE CON SCALA E SPECIFICARE MOTIVI ASSENZA REQUISITI | | | | | |
|---|---|---|---|---|---|
| **SCALE FROM 1 TO 5 (FOR ALL REQUIREMENTS)** | | | | | |
| **PRELIMINARY REQUIREMENTS** | **1**<br>**Almost absent** | **2** | **3** | **4** | **5**<br>**Fully present** |
| ☐ PATIENT SAFETY | 1<br>2<br>3<br>4<br>5 | | | | |
| ☐ OPERATOR SAFETY | 1<br>2<br>3<br>4<br>5 | | | | |
| ☐ PATIENT COMFORT | 1<br>2<br>3<br>4<br>5 | | | | |
| ☐ LOW APPLIED PHYSICAL EFFORT | 1<br>2<br>3<br>4<br>5 | | | | |
| ☐ EASY TO USE | 1<br>2<br>3<br>4<br>5 | | | | |
| **SPECIFIC REQUIREMENTS** | **1**<br>**Almost absent** | **2** | **3** | **4** | **5**<br>**Fully present** |
| ☐ ADAPTABLE TO PATIENT | 1<br>2<br>3<br>4<br>5 | | | | |
| ☐ ADAPTABLE TO TYPE OF HANDLING | 1<br>2<br>3<br>4<br>5 | | | | |
| ☐ ADAPTABLE TO ENVIRONMENT | 1<br>2<br>3<br>4<br>5 | | | | |

**FIGURE 10.1** _(continued)_    Ward sheet choice aids. _(continued)_

**THIRD STEP: FINAL SUMMARY AFTER DELIVERY OF EQUIPMENT**

| MANUAL HANDLING: describe routine tasks involving total or partial patient lifting | Total lifting (TL) WITHOUT EQUIPMENT | | | Partial Lifting (PL) WITHOUT EQUIPMENT | | |
|---|---|---|---|---|---|---|
| indicate the **nnumber of tasks** per shift involving manual patient handling | morning | afternoon | night | morning | afternoon | night |
| | A | B | C | D | E | F |
| ☐ pulling up in bed | ☐☐☐☐ | ☐☐☐☐ | ☐☐☐☐ | ☐☐☐☐ | ☐☐☐☐ | ☐☐☐☐ |
| ☐ turning over in bed (to change position) | | | | ☐☐☐☐☐ | ☐☐☐☐☐ | ☐☐☐☐☐ |
| ☐ bed-to-wheelchair and viceversa | ☐☐ | ☐☐ | ☐☐ | ☐☐ | ☐☐ | ☐☐ |
| ☐ lifting from seated to upright position | | | | ☐☐ | ☐☐ | ☐☐ |
| ☐ bed-to-stretcher and viceversa | ☐☐ | ☐☐ | ☐☐ | ☐☐ | ☐☐ | ☐☐ |
| ☐ wheelchair-to-toilet and viceversa | ☐☐ | ☐☐ | ☐☐ | ☐☐ | ☐☐ | ☐☐ |
| ☐ other | ☐☐ | ☐☐ | ☐☐ | ☐☐ | ☐☐ | ☐☐ |
| ☐ other | ☐☐ | ☐☐ | ☐☐ | ☐☐ | ☐☐ | ☐☐ |
| TOTAL: calculate the total for each column | | | | | | |
| **Number of total (TL) or partial (PL) manual lifting tasks** | A+B+C = TL | | | D+E+F=PL | | |

| AIDED HANDLING: describe routine tasks involving total or partial patients lifting using available equipment | Total lifting (TL) AIDED | | | Partial Lifting (PL) AIDED | | |
|---|---|---|---|---|---|---|
| indicate the number of tasks per shift involving aided patient handling | morning | afternoon | night | morning | afternoon | night |
| | G | H | I | L | M | N |
| ☐ pulling up in bed | ☐☐☐☐ | ☐☐☐☐ | ☐☐☐☐ | ☐☐☐☐ | ☐☐☐☐ | ☐☐☐☐ |
| ☐ turning over in bed (reposition) | | | | ☐☐☐☐☐ | ☐☐☐☐☐ | ☐☐☐☐☐ |
| ☐ bed/wheelchair | ☐☐ | ☐☐ | ☐☐ | ☐☐ | ☐☐ | ☐☐ |
| ☐ lifting from seated to upright position | | | | ☐☐ | ☐☐ | ☐☐ |
| ☐ bed/stretcher | ☐☐ | ☐☐ | ☐☐ | ☐☐ | ☐☐ | ☐☐ |
| ☐ wheelchair/toilet | ☐☐ | ☐☐ | ☐☐ | ☐☐ | ☐☐ | ☐☐ |
| ☐ other | ☐☐ | ☐☐ | ☐☐ | ☐☐ | ☐☐ | ☐☐ |
| ☐ other | ☐☐ | ☐☐ | ☐☐ | ☐☐ | ☐☐ | ☐☐ |
| TOTAL: calculate the total for each column | | | | | | |
| **AIDED handling total (ATL) or partial (APL) lifting** | G+H+I = ATL | | | L+M+N=APL | | |
| **% OF AIDED TOTAL LIFTING OPERATIONS (% ATL)** | ATL (TL + ATL) | | | | | |
| **% OF AIDED PARTIAL LIFTING OPERATIONS (% APL)** | | | | APL (PL + APL) | | |

$$\text{MAPO INDEX} = (|\underline{\quad}| \times |\underline{\quad}| + |\underline{\quad}| \times |\underline{\quad}|) \times |\underline{\quad}| \times |\underline{\quad}| \times |\underline{\quad}| =$$
$$\phantom{MAPO} \quad NC/OP \quad LF \quad PC/OP \quad AF \quad WF \quad EF \quad TF$$

**FIGURE 10.1** *(continued)*    Ward sheet choice aids.

FONDAZIONE IRCCS CA' GRANDA
Ospedale Maggiore Policlinico
CLINICA DEL LAVORO – MILAN ITALY
ERGONOMICS SECTION

## WARDS: DATA COLLECTION SHEET FOR CHOICE OF AIDS

1. DATA FROM WARD DATA COLLECTION SHEET

| HOSPITAL: | WARD: | | WARD CODE: |
|---|---|---|---|
| Nr BEDS: | AVERAGE HOSPITAL STAY (days): | | DATE: |
| **% OF AIDED TOTAL LIFTING OPERATIONS (%ATL)** | | | |
| **% OF AIDED PARTIAL LIFTING OPERATIONS (%APL)** | | | |

MAPO = (|_____| x |____|+ |_____| x |____|) x |____| x|_____| x |____| =

INDEX    NC/OP    LF    PC/OP    AF    WF    EF    TF

| MANUAL PATIENT HANDLING TASKS | | | | | | |
|---|---|---|---|---|---|---|
| **MANUAL HANDLING:** describe routine tasks involving total or partial patient lifting | **Total lifting (TL)** WITHOUT EQUIPMENT | | | **Partial Lifting (PL)** WITHOUT EQUIPMENT | | |
| indicate the **number of tasks** per shift involving manual patient handling | morning | afternoon | night | morning | afternoon | night |
| | A | B | C | D | E | F |
| ☐ **pulling up in bed** | ☐☐☐☐ | ☐☐☐☐ | ☐☐☐☐ | ☐☐☐☐ | ☐☐☐☐ | ☐☐☐☐ |
| ☐ **turning over in bed (to change position)** | | | | ☐☐☐☐☐ | ☐☐☐☐☐ | ☐☐☐☐☐ |
| ☐ **bed-to-wheelchair and viceversa** | ☐☐ | ☐☐ | ☐☐ | ☐☐ | ☐☐ | ☐☐ |
| ☐ **lifting from seated to upright position** | | | | ☐☐ | ☐☐ | ☐☐ |
| ☐ **bed-to-stretcher and viceversa** | ☐☐ | ☐☐ | ☐☐ | ☐☐ | ☐☐ | ☐☐ |
| ☐ **wheelchair-to-toilet and viceversa** | ☐☐ | ☐☐ | ☐☐ | ☐☐ | ☐☐ | ☐☐ |
| ☐ **other** | ☐☐ | ☐☐ | ☐☐ | ☐☐ | ☐☐ | ☐☐ |
| ☐ **other** | ☐☐ | ☐☐ | ☐☐ | ☐☐ | ☐☐ | ☐☐ |
| **TOTAL:** calculate the total for each column | | | | | | |
| **Number of total (TL) or partial (PL) manual lifting tasks** | A+B+C = TL | | | D+E+F=PL | | |

### FIRST STEP: CHOICE OF AIDS FOR RISK REDUCTION

| | |
|---|---|
| ☐ ADJUSTABLE STRETCHERS (Nr = |__|__| ) | ☐ SLIDING BOARDS (Nr |__|__| ) |
| **REQUIREMENTS OF HOISTS :** | **TYPE AND NUMBER OF SLINGS :** |
| ☐ CEILING MOUNTED HOIST     Nr = |__|__| | _____ |
| | _____ |
| ☐ WHEELED HOIST     N°= |__|__| | TYPE 1 Nr = |__|TYPE 2 Nr = |__|TYPE 2 Nr = |__| |
| ☐ STANDING HOIST     N°= |__|__| | |
| ☐ ADJUSTABLE AND ERGONOMIC BEDS   (N°= |__|__| ) | |
| ☐ SLIDING SHEETS (N°= |__|__| )     SIZE _____     TYPE _____ | |
| ☐ ERGONOMIC BELTS (N°= |__|__| ) SIZE _____     TYPE _____ | |
| OTHER: _____ _____(N°= |__|__| ) | |
| ☐ BATHING STRETCHER | ☐ HEIGHT ADJUSTABLE BATH |
| ☐ BATHING CHAIR | ☐ SHOWER FLOOR |
| ☐ WHEELCHAIRS     TYPE_____     (N°= |__|__| ) | |

**FIGURE 10.2**   Surgical unit choice of aids sheet. *(continued)*

**SECOND STEP: HOW TO ASSESS ERGONOMIC REQUIREMENTS OF AIDS**

**TYPE OF EQUIPMENT:** _____

| PRELIMINARY REQUIREMENTS: VALUTARE CON SCALA E SPECIFICARE MOTIVI ASSENZA REQUISITI | | | | | |
|---|---|---|---|---|---|
| SCALE FROM 1 TO 5 (FOR ALL REQUIREMENTS) | | | | | |
| **PRELIMINARY REQUIREMENTS** | **1**<br>**Almost absent** | **2** | **3** | **4** | **5**<br>**Fully present** |
| ☐ PATIENT SAFETY | 1<br>2<br>3<br>4<br>5 | | | | |
| ☐ OPERATOR SAFETY | 1<br>2<br>3<br>4<br>5 | | | | |
| ☐ PATIENT COMFORT | 1<br>2<br>3<br>4<br>5 | | | | |
| ☐ LOW APPLIED PHYSICAL EFFORT | 1<br>2<br>3<br>4<br>5 | | | | |
| ☐ EASY TO USE | 1<br>2<br>3<br>4<br>5 | | | | |
| **SPECIFIC REQUIREMENTS** | **1**<br>**Almost absent** | **2** | **3** | **4** | **5**<br>**Fully present** |
| ☐ ADAPTABLE TO PATIENT | 1<br>2<br>3<br>4<br>5 | | | | |
| ☐ ADAPTABLE TO TYPE OF HANDLING | 1<br>2<br>3<br>4<br>5 | | | | |
| ☐ ADAPTABLE TO ENVIRONMENT | 1<br>2<br>3<br>4<br>5 | | | | |

**FIGURE 10.2** *(continued)*     Surgical unit choice of aids sheet. *(continued)*

FONDAZIONE IRCCS CA' GRANDA
OSPEDALE MAGGIORE POLICLINICO
CLINICA DEL LAVORO – MILAN (ITALY)
ERGONOMICS SECTION

**THIRD STEP: FINAL SUMMARY AFTER DELIVERY OF EQUIPMENT**

| MANUAL HANDLING: describe routine tasks involving total or partial patient lifting | Total lifting (TL) WITHOUT EQUIPMENT | | | Partial Lifting (PL) WITHOUT EQUIPMENT | | |
|---|---|---|---|---|---|---|
| indicate the **number of tasks** per shift involving manual patient handling | morning | afternoon | night | morning | afternoon | night |
| | A | B | C | D | E | F |
| ☐ pulling up in bed | ☐☐☐☐ | ☐☐☐☐ | ☐☐☐☐ | ☐☐☐☐ | ☐☐☐☐ | ☐☐☐☐ |
| ☐ turning over in bed (to change position) | | | | ☐☐☐☐☐ | ☐☐☐☐☐ | ☐☐☐☐☐ |
| ☐ bed-to-wheelchair and viceversa | ☐☐ | ☐☐ | ☐☐ | ☐☐ | ☐☐ | ☐☐ |
| ☐ lifting from seated to upright position | | | | ☐☐ | ☐☐ | ☐☐ |
| ☐ bed-to-stretcher and viceversa | ☐☐ | ☐☐ | ☐☐ | ☐☐ | ☐☐ | ☐☐ |
| ☐ wheelchair-to-toilet and viceversa | ☐☐ | ☐☐ | ☐☐ | ☐☐ | ☐☐ | ☐☐ |
| ☐ other | ☐☐ | ☐☐ | ☐☐ | ☐☐ | ☐☐ | ☐☐ |
| ☐ other | ☐☐ | ☐☐ | ☐☐ | ☐☐ | ☐☐ | ☐☐ |
| TOTAL: calculate the total for each column | | | | | | |
| **Number of total (TL) or partial (PL) manual lifting tasks** | A+B+C = TL | | | D+E+F=PL | | |

| AIDED HANDLING: describe routine tasks involving total or partial patients lifting using available equipment | Total lifting (TL) AIDED | | | Partial Lifting (PL) AIDED | | |
|---|---|---|---|---|---|---|
| Indicate the number of tasks per shift involving aided patient handling | morning | afternoon | night | morning | afternoon | night |
| | G | H | I | L | M | N |
| ☐ pulling up in bed | ☐☐☐☐ | ☐☐☐☐ | ☐☐☐☐ | ☐☐☐☐ | ☐☐☐☐ | ☐☐☐☐ |
| ☐ turning over in bed (reposition) | | | | ☐☐☐☐☐ | ☐☐☐☐☐ | ☐☐☐☐☐ |
| ☐ bed/wheelchair | ☐☐ | ☐☐ | ☐☐ | ☐☐ | ☐☐ | ☐☐ |
| ☐ lifting from seated to upright position | | | | ☐☐ | ☐☐ | ☐☐ |
| ☐ bed/stretcher | ☐☐ | ☐☐ | ☐☐ | ☐☐ | ☐☐ | ☐☐ |
| ☐ wheelchair/toilet | ☐☐ | ☐☐ | ☐☐ | ☐☐ | ☐☐ | ☐☐ |
| ☐ other | ☐☐ | ☐☐ | ☐☐ | ☐☐ | ☐☐ | ☐☐ |
| ☐ other | ☐☐ | ☐☐ | ☐☐ | ☐☐ | ☐☐ | ☐☐ |
| TOTAL: calculate the total for each column | | | | | | |
| **AIDED handling total (ATL) or partial (APL) lifting** | G+H+I = ATL | | | L+M+N=APL | | |
| **% OF AIDED TOTAL LIFTING OPERATIONS (% ATL)** | ATL (TL + ATL) | | | | | |
| **% OF AIDED PARTIAL LIFTING OPERATIONS (% APL)** | | | | APL (PL + APL) | | |

MAPO = (|_____| x |_____|+ |_____| x |____|) x |____| x |_____| x |____| =
INDEX    NC/OP   LF    PC/OP    AF    WF    EF    TF

**FIGURE 10.2** *(continued)* Surgical unit choice of aids sheet.

## PATIENT HANDLING RISK IN <u>SURGICAL UNIT</u>: CORRECTIVE ACTIONS INVOLVING LIFTING EQUIPMENT

### 1. DATA FROM WARD DATA COLLECTION SHEET

| HOSPITAL : | WARD : | | WARD CODE : | |
|---|---|---|---|---|
| Nr BEDS : | AVERAGE HOSPITAL STAY (days) : | | DATE : | |
| **% OF AIDED TOTAL LIFTING OPERATIONS (% ATL)** | | | | |
| **% OF AIDED PARTIAL LIFTING OPERATIONS (% APL)** | | | | |

MAPO = (|_____| x |____|+ |_____| x |____|) x |____| x|_____| x |____| =

INDEX     NC/OP     LF     PC/OP     AF     WF     EF     TF

| **MANUAL PATIENT HANDLING TASKS** | | | | | | |
|---|---|---|---|---|---|---|
| **MANUAL HANDLING:** describe routine tasks involving total or partial patient lifting | **Total lifting (TL) WITHOUT EQUIPMENT** | | | **Partial Lifting (PL) WITHOUT EQUIPMENT** | | |
| indicate the **number of tasks** per shift involving manual patient handling | morning | afternoon | night | morning | afternoon | night |
| | A | B | C | D | E | F |
| ☐ pulling up in bed | ☐☐☐☐ | ☐☐☐☐ | ☐☐☐☐ | ☐☐☐☐ | ☐☐☐☐ | ☐☐☐☐ |
| ☐ turning over in bed (to change position) | | | | ☐☐☐☐☐ | ☐☐☐☐☐ | ☐☐☐☐☐ |
| ☐ bed-to-wheelchair and viceversa | ☐☐ | ☐☐ | ☐☐ | ☐☐ | ☐☐ | ☐☐ |
| ☐ lifting from seated to upright position | | | | ☐☐ | ☐☐ | ☐☐ |
| ☐ bed-to-stretcher and viceversa | ☐☐ | ☐☐ | ☐☐ | ☐☐ | ☐☐ | ☐☐ |
| ☐ wheelchair-to-toilet and viceversa | ☐☐ | ☐☐ | ☐☐ | ☐☐ | ☐☐ | ☐☐ |
| ☐ other | ☐☐ | ☐☐ | ☐☐ | ☐☐ | ☐☐ | ☐☐ |
| ☐ other | ☐☐ | ☐☐ | ☐☐ | ☐☐ | ☐☐ | ☐☐ |
| **TOTAL:** calculate the total for each column | | | | | | |
| **Number of total (TL) or partial (PL) manual lifting tasks** | A+B+C = TL | | | D+E+F=PL | | |

<u>FIRST STEP: CHOICE OF AIDS FOR RISK REDUCTION</u>

| ☐ ADJUSTABLE STRETCHERS (Nr = |__|__| ) | ☐ SLIDING BOARDS (Nr |__|__| ) |
|---|---|
| **REQUIREMENTS OF HOISTS :** | **TYPE AND NUMBER OF SLINGS :** |
| ☐ CEILING MOUNTED HOIST      Nr = |__|__| | ---------------------- ---------------------- |
| ☐ WHEELED HOIST      N°= |__|__| | TYPE 1 Nr = |__|TYPE 2 Nr = |__|TYPE 2 Nr = |__| |
| ☐ STANDING HOIST      N°= |__|__| | |
| ☐ ADJUSTABLE AND ERGONOMIC BEDS      (N°= |__|__| ) | |
| ☐ SLIDING SHEETS (N°= |__|__| )      SIZE _____      TYPE _____ | |
| ☐ ERGONOMIC BELTS (N°= |__|__| )      SIZE _____      TYPE _____ | |
| OTHER: _____ _____(N°= |__|__| ) | |
| ☐ BATHING STRETCHER | ☐ HEIGHT ADJUSTABLE BATH |
| ☐ BATHING CHAIR | ☐ SHOWER FLOOR |
| ☐ WHEELCHAIRS      TYPE_____ | (N°= |__|__| ) |

**FIGURE 10.3**   Outpatient Services choice of aids sheet. *(continued)*

**SECOND STEP: HOW TO ASSESS ERGONOMIC REQUIREMENTS OF AIDS**

**TYPE OF EQUIPMENT:**

| PRELIMINARY REQUIREMENTS: VALUTARE CON SCALA E SPECIFICARE MOTIVI ASSENZA REQUISITI | | | | | |
|---|---|---|---|---|---|
| **SCALE FROM 1 TO 5 (FOR ALL REQUIREMENTS)** | | | | | |
| **PRELIMINARY REQUIREMENTS** | **1** <br> **Almost absent** | **2** | **3** | **4** | **5** <br> **Fully present** |
| ☐ PATIENT SAFETY | 1 <br> 2 <br> 3 <br> 4 <br> 5 | | | | |
| ☐ OPERATOR SAFETY | 1 <br> 2 <br> 3 <br> 4 <br> 5 | | | | |
| ☐ PATIENT COMFORT | 1 <br> 2 <br> 3 <br> 4 <br> 5 | | | | |
| ☐ LOW APPLIED PHYSICAL EFFORT | 1 <br> 2 <br> 3 <br> 4 <br> 5 | | | | |
| ☐ EASY TO USE | 1 <br> 2 <br> 3 <br> 4 <br> 5 | | | | |
| **SPECIFIC REQUIREMENTS** | **1** <br> **Almost absent** | **2** | **3** | **4** | **5** <br> **Fully present** |
| ☐ ADAPTABLE TO PATIENT | 1 <br> 2 <br> 3 <br> 4 <br> 5 | | | | |
| ☐ ADAPTABLE TO TYPE OF HANDLING | 1 <br> 2 <br> 3 <br> 4 <br> 5 | | | | |
| ☐ ADAPTABLE TO ENVIRONMENT | 1 <br> 2 <br> 3 <br> 4 <br> 5 | | | | |

**FIGURE 10.3** *(continued)* Outpatient Services choice of aids sheet. *(continued)*

**THIRD STEP: FINAL SUMMARY AFTER DELIVERY OF EQUIPMENT**

| MANUAL HANDLING: describe routine tasks involving total or partial patient lifting | Total lifting (TL) WITHOUT EQUIPMENT | | | Partial Lifting (PL) WITHOUT EQUIPMENT | | |
|---|---|---|---|---|---|---|
| indicate the **number of tasks** per shift involving manual patient handling | morning | afternoon | night | morning | afternoon | night |
| | **A** | **B** | **C** | **D** | **E** | **F** |
| ☐ pulling up in bed | ☐☐☐☐ | ☐☐☐☐ | ☐☐☐☐ | ☐☐☐☐ | ☐☐☐☐ | ☐☐☐☐ |
| ☐ turning over in bed (to change position) | | | | ☐☐☐☐☐ | ☐☐☐☐☐ | ☐☐☐☐☐ |
| ☐ bed-to-wheelchair and viceversa | ☐☐ | ☐☐ | ☐☐ | ☐☐ | ☐☐ | ☐☐ |
| ☐ lifting from seated to upright position | | | | ☐☐ | ☐☐ | ☐☐ |
| ☐ bed-to-stretcher and viceversa | ☐☐ | ☐☐ | ☐☐ | ☐☐ | ☐☐ | ☐☐ |
| ☐ wheelchair-to-toilet and viceversa | ☐☐ | ☐☐ | ☐☐ | ☐☐ | ☐☐ | ☐☐ |
| ☐ other | ☐☐ | ☐☐ | ☐☐ | ☐☐ | ☐☐ | ☐☐ |
| ☐ other | ☐☐ | ☐☐ | ☐☐ | ☐☐ | ☐☐ | ☐☐ |
| **TOTAL:** calculate the total for each column | | | | | | |
| **Number of total (TL) or partial (PL) manual lifting tasks** | A+B+C = TL | | | D+E+F=PL | | |

| AIDED HANDLING: describe routine tasks involving total or partial patients lifting using available equipment | Total lifting (TL) AIDED | | | Partial Lifting (PL) AIDED | | |
|---|---|---|---|---|---|---|
| indicate the number of tasks per shift involving aided patient handling | morning | afternoon | night | morning | afternoon | night |
| | **G** | **H** | **I** | **L** | **M** | **N** |
| ☐ pulling up in bed | ☐☐☐☐ | ☐☐☐☐ | ☐☐☐☐ | ☐☐☐☐ | ☐☐☐☐ | ☐☐☐☐ |
| ☐ turning over in bed (reposition) | | | | ☐☐☐☐☐ | ☐☐☐☐☐ | ☐☐☐☐☐ |
| ☐ bed/wheelchair | ☐☐ | ☐☐ | ☐☐ | ☐☐ | ☐☐ | ☐☐ |
| ☐ lifting from seated to upright position | | | | ☐☐ | ☐☐ | ☐☐ |
| ☐ bed/stretcher | ☐☐ | ☐☐ | ☐☐ | ☐☐ | ☐☐ | ☐☐ |
| ☐ wheelchair/toilet | ☐☐ | ☐☐ | ☐☐ | ☐☐ | ☐☐ | ☐☐ |
| ☐ other | ☐☐ | ☐☐ | ☐☐ | ☐☐ | ☐☐ | ☐☐ |
| ☐ other | ☐☐ | ☐☐ | ☐☐ | ☐☐ | ☐☐ | ☐☐ |
| **TOTAL:** calculate the total for each column | | | | | | |
| **AIDED handling total (ATL) or partial (APL) lifting** | G+H+I = ATL | | | L+M+N=APL | | |
| **% OF AIDED TOTAL LIFTING OPERATIONS (% ATL)** | ATL (TL + ATL) | | | | | |
| **% OF AIDED PARTIAL LIFTING OPERATIONS (% APL)** | | | | APL (PL + APL) | | |

MAPO    = (|_____| x |_____|+ |_____| x |_____|) x |____| x |____| x |____| =
INDEX        NC/OP    LF      PC/OP    AF        WF        EF        TF

**FIGURE 10.3** *(continued)*   Outpatient Services choice of aids sheet.

# 11 Definition and Management of Preventive Planning

## 11.1 INTRODUCTION

The international scientific community is currently emphasizing the importance of adopting *multifactorial* approaches and strategies for the prevention of musculo-skeletal damage due to exposure to manual patient-handling risk; such approaches are based on ergonomic principles (both top–down and bottom–up, and especially on the engagement of all the various parties involved).

While the decision as to how risk should be controlled depends largely on national regulations, the following aspects should also be taken into account:

- Creation of a team of professionals tasked with managing specific manual patient-handling (MPH) risk, reducing manual patient-handling risk by adopting a complex multifactorial strategy encompassing targeted, modern-day, coordinated actions beginning with the introduction of adequate patient-handling aids and including training staff to use the equipment properly and correctly move patients manually, as well as reorganizing work and procedures and redesigning work environments (Hignett 2003). It is essential for management to approve and support such interventions formally, and for every worker's role and responsibilities to be defined and communicated clearly.
- Establishment within the healthcare facility of "back-care advisors" (variously called a peer leader, ergoranger, or ergocoach)—that is, a full-time ergonomics expert responsible for the prevention of manual patient-handling risk (Nelson et al. 2006; Knibbe and Knibbe 2006)
- Identification of specific procedures to be followed for acquiring patient moving aids and equipment. The use of policies and procedures helps the organization to direct its resources and personnel to behave in a positive fashion to the potential problems. Studies have reported positive effects of the implementation of policies and procedures to assist with the implementation of best practices and the reduction of organizational losses (Garg 2006; Collins et al. 2004; Yassi et al. 2001).
- Establishment of a "lifelong learning" group
- Creation or, if present, involvement of the occupational health service
- Continuous monitoring of the effectiveness of risk prevention strategies
- National regulators to encourage the definition of application rules for the reduction of manual patient-handling risk

## 11.2  APPROACHES TO STAFF TRAINING

A strategic plan to contain and generally manage manual patient-handling risk consists of delivering information and training as part of a broader preventive approach encompassing short-term corrective actions and the provision of equipment suitable for both the patient and the operator.

More specifically, **information** can be defined as the "targeted" delivery to all the subjects involved of news, updates, and facts upon which a more comprehensive prevention process will be based. **Training** means the acquisition of education by the subjects involved, as well as the appropriate operating and behavioral skills to drive the adoption of new ways of "thinking and acting" with respect to safety, by changing habitual behaviors (which may entail using patient lifting or moving aids, for instance) so as to put in practice the rules and principles underlying workplace health and safety, as well as to prevent risk and handle emergencies.

Countless studies have been carried out since the late 1980s on minimizing spinal damage solely by adopting appropriate patient-lifting techniques (Videman et al. 1989); as emphasized in a 2003 review by Hignett, these have been proven to be largely **ineffective** for significantly reducing both the level of the compressive forces acting on intervertebral disks and the level of low-back pain.

Only the proper utilization of adequate equipment can contribute to lowering risk for the lumbar spine significantly (Burdof, Koppelaar, and Bradley 2013). Generally speaking, training for adult operators should follow the lifelong learning principles developed by Malcom Knowles (Knowles, Holton, and Swanson 2008).

The aim of training is to engage the largest possible number of individuals exposed to the specific risk factors and provide them with a **specific understanding** of the characteristics and extent of the relevant **risk factors** and their mode of action, and of how individuals and collective groups should behave in order to make proper use of the lifting devices and equipment provided in different situations (procedures and algorithms).

Based on the training experiences mentioned in the introduction to this chapter and on the preventive strategies put in place and effectiveness tested by countless hospitals over the past 10 years, several different levels of training may be implemented:

1. Theoretical and practical courses delivered by experts (occupational physiotherapists) to groups of operators
2. Creation of a group of trainers (physiotherapists and senior ward staff) to deliver lifelong learning/continuous training based on an agenda of clarifying objectives, setting training priorities, defining methods and processes for transferring knowledge, and formulating corporate strategies; the "lifelong learning" (continuous training) team must be formally recognized by hospital management and will negotiate the setting within which training is to be delivered
3. Appointment of ergocoaches (or peer leaders), from among the ward staff, to be trained specifically in ergonomics; the underlying goal is to have an

in-house operator capable of monitoring staff during their shifts and solving any patient mobilization issues

4. Specific training regarding university courses in nursing sciences

In the Dutch experience (Knibbe and Knibbe 2006) as well as in our own experience in Italy, the most successful results have been achieved by training trainers (ergocoaches or lifelong learning/continuous training teams), but this approach requires enough staff to allocate specifically to this purpose.

A multifactorial approach means that even macroergonomic interventions are a necessary step toward bringing together all the various elements that contribute to tailoring patient care to individual needs and designing a preventive strategy cutting across all hospital departments and services (i.e., prevention services, management, equipment procurement, etc.).

Effectiveness testing is a critical part of all occupational risk prevention strategies.

## 11.3  TESTING THE EFFECTIVENESS OF PREVENTIVE STRATEGIES

Effectiveness testing is all the more essential at times like this, when the best possible results must be squeezed out of limited financial resources. Preventive strategies obviously require investments and thus constitute a cost; however, it should be emphasized that the lack of prevention is also a cost.

In summary, based on evidence drawn from the literature (Mastrangelo 2008; Hashemi, Webster, and Clancy 1998), the cost of nonprevention (both direct and indirect) lies in occupational diseases and disorders, the in-house management of accidents and injuries caused by biomechanical overload, insurance premiums, healthcare spending on tests and analyses, time off work due to sickness among healthcare workers, additional training for people transferred to other duties, and so on. By contrast, the cost of prevention includes risk assessments, the purchase of lifting devices and aids, personnel training, etc.

As a general rule, effectiveness training should envisage the use of process and performance indicators. Process indicators include various tools for assessing postures adopted during specific lifting maneuvers: REBA (Hignett and McAtamney 2000) and OWAS (Karhu, Kansi, and Kuorinka 1977). These methods involve the examination of individual frames from a film, chosen on the basis of questionable criteria, to analyze the postures adopted by various body parts and assigning scores ranging from high to low using a ranking scale to indicate the presence of risk.

The observational tools developed to assess the safety and skill of operators involved in manual patient-handling activities are also worth mentioning: simple checklists have been developed, for example, by Feldstein, Vollmer, and Valanis (1990), Kjellberg et al. (2000), St. Vincent, Tellier, and Lortie (1989), and Engels et al. (1997). A few are also mentioned in technical report (TR) 12296: PATE (Kjellberg et al. 2000) and DiNO (Johnsson et al. 2004).

### 11.3.1 Mapping Risk in a Hospital: Examples of Procedures and Applications

The risk analysis method proposed in this book is a tool for evaluating processes in terms of defining risk indexes as well as individual risk factors. It is actually possible to track changes in individual risk factors over time, even if the risk level remains the same. The percentage of patient-handling tasks performed using aids is a good example of this.

Following is a selection of tools that can be used during the analytical phase to map risk in the areas of a hospital where up to 90% of the staff could be exposed to manual patient-handling risk (i.e., patient wards).

In healthcare facilities where operators are exposed to patient-handling risk, risk must be mapped in order to identify where risk levels are the highest, to plan corrective actions to reduce risk, and to better manage staff in terms of workloads and taking into account staff exempted from performing certain tasks.

Another way to consider the extent of risk to which workers are exposed is the incidence of on-the-job injuries, especially musculoskeletal damage caused by patient-care and -handling activities.

Risk mapping can be used to manage a prevention plan linking the two aspects, covering

- Ward-specific organizational aspects (total staff numbers, number of beds, number of disabled or noncooperative patients, etc.)
- The various risk factors making up the MAPO (*movimentazione e assistenza pazienti ospedalizzati*—movement and assistance for hospitalized patients) index and consequent level of exposure to patient-handling risk
- The percentage of staff members with musculoskeletal injuries
- An overall summary of the hospital units (or wards) analyzed

A software program has been designed for this specific purpose that can be downloaded for free from the EPM website: http://www.epmresearch.org/userfiles/files/mapping%20MAPO%202014.xls (file "mapping MAPO 2014.xls") and consists of several sections:

- **Section 1:** One page containing the results of the risk assessments in all units analyzed ("hospital" sheet). Figure 11.1 provides a sample data sheet showing that in this particular case, risk assessments have been carried out in only ten wards (45% of the total). The other data entered into the form refer to general characteristics of the hospital that might potentially be linked to patient-handling risk.
- **Section 2:** One page to be used for filing both the results of the MAPO risk assessment and of MPH-related injuries ("wards" sheet).
- **Section 3:** One page with a graphic depiction of the data collected in the previous sheets.

The first page is for entering general data concerning the hospital or healthcare facility to be assessed, in particular the total number of wards and the year of the risk assessment. The other fields are populated automatically when data concerning the risk assessments of the various units are entered into the following pages. This generates an up-to-the-minute picture of the status of the assessment process within the hospital or facility.

The results of the analytical evaluation are then entered into the next section. There is a line for each separate data sheet. The data on any patient-handling injuries occurring in the ward are also entered on the same line.

The focus is on low-back injuries relating to manual patient-handling activities.

The parameters used in the "wards" sheet for ten wards are listed in Figures 11.2 and 11.3, especially the data concerning

- Staff engaged in MPH
- Risk factors (e.g., each risk factor score is "translated" into red if completely inadequate, yellow if partially inadequate, and green if adequate), along with the percentage of aided patient-handling tasks
- Patient-handling-related injuries in each of the wards analyzed

The program automatically calculates the exposure level in the column under the heading "MAPO index" and shows levels in different colors.

All the data collected are processed and presented as results in the "summary and graphs" section: distribution of the various units based on their exposure level, adequacy of the individual risk factors, distribution of staff based on exposure level (Figures 11.4, 11.5, and 11.6). Figure 11.7 shows the data relating to MPH injuries, which are an additional aspect pertaining to the specific risk.

## 11.4 OTHER EFFECTIVENESS-TESTING TOOLS

Subjective evaluations by staff and patients represent poorly developed yet valid tools that can be used to test effectiveness. For example, for staff, the Borg scale (Borg 1998) can be used to assess perceived effort during patient-handling tasks, while for patients, comfort during handling (Nelson et al. 2006), perceived safety, and respect for patient privacy may be regarded as good examples, although specific tools are not yet available.

It is also worth mentioning learning processes relating to specific risk training, in addition to those already referred to or described in previous chapters. But instead of focusing on the tools, attention should focus on the resources employed for this purpose. There is substantial evidence (Knibbe et al. 2008) pointing to the need to distribute both training and effectiveness testing over a wider number of subjects—hence, the recommendation to make every effort to ensure that each ward or unit has a person on hand who has been specifically trained in MPH risk (ergocoach); this approach is also encouraged in TR 12296.

Some of the most widely utilized effectiveness-testing tools include assessments of trends in patient-handling-related injuries, occupational diseases, disorders due to biomechanical overload, and the periodic monitoring of musculoskeletal disorders in units or wards in which a health surveillance system has been implemented by occupational health specialists (Fray 2010).

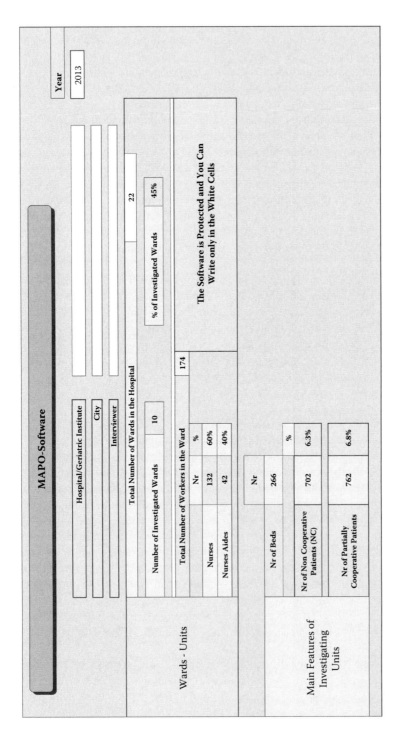

**FIGURE 11.1**  Hospital sheet.

| | | RISK MAPPING | | | | | | | | | | | | | | | | |
|---|---|---|---|---|---|---|---|---|---|---|---|---|---|---|---|---|---|---|
| | | Staff engaged in MPH | | | Multiplying Factors | | | | | | | | | | | | | |
| **WARD CODE** | **WARD - UNIT (MANDATORY FIELD)** | **Nr of Nurses** | **Nr of Nurses Aides** | **Total number of exposed workers** | **Nr of Beds** | **NC** (Non Cooperative patients) | **OP** (Operators engaged in MPH in 24 hours) | **PC** (Partially Cooperative patients) | **NC/OP** | **PC/OP** | % of aided total lifting operations **(%ATL)** | **LF** (lifting device factor) | % of aided partial lifting operations **(%APL)** | **AF** (minor aids factor) | **WF** (wheelchair factor) | **EF** (environment factor) | **TF** (training factor) | **MAPO Index** |
| 15 | MEDICINE 1 | 8 | 7 | 15 | **22** | 8 | 11 | 7 | 0,7 | 0,6 | **37%** | 2,00 | **44%** | 1,00 | 1,12 | 1,25 | 2,00 | 5,58 |
| 1 | SURGERY 1 | 10 | 4 | 14 | **20** | 5 | 8 | 3 | 0,6 | 0,4 | **30%** | 2,00 | **0%** | 1,00 | 0,75 | 0,75 | 0,75 | **0,69** |
| 33 | GERIATRIC | 10 | 8 | 18 | **34** | 9 | 10 | 8 | 0,9 | 0,8 | **0%** | 4,00 | **25%** | 1,00 | 1,50 | 1,25 | 1,00 | 8,25 |
| 2 | CARDIOLOGIC | 10 | 8 | 18 | **30** | 4 | 11 | 9 | 0,4 | 0,8 | **0%** | 4,00 | **50%** | 1,00 | 0,75 | 0,75 | 2,00 | **2,56** |
| 3 | SURGERY 2 | 14 | 4 | 18 | **20** | 6 | 8 | 3 | 0,8 | 0,4 | **40%** | 2,00 | **90%** | 0,50 | 1,00 | 1,25 | 1,00 | **2,11** |
| 32 | MEDICINE 2 | 13 | 0 | 13 | **31** | 7 | 9 | 15 | 0,8 | 1,7 | **42%** | 2,00 | **0%** | 1,00 | 1,00 | 1,25 | 2,00 | 8,06 |
| 11 | GYNECOLOGY | 16 | 0 | 16 | **26** | 0 | 7 | 12 | 0,0 | 1,7 | **0%** | 0,50 | **0%** | 1,00 | 0,75 | 1,25 | 2,00 | **3,21** |
| 21 | ORTHOPEDY | 18 | 4 | 22 | **32** | 19 | 11 | 6 | 1,7 | 0,5 | **0%** | 4,00 | **0%** | 1,00 | 1,50 | 1,50 | 1,00 | 16,77 |
| 13 | GASTROEN-TEROLOGY | 13 | 3 | 16 | **20** | 2 | 8 | 3 | 0,3 | 0,4 | **8%** | 2,00 | **20%** | 1,00 | 0,75 | 1,25 | 2,00 | **1,64** |
| 38 | NEUROLOGY | 20 | 4 | 24 | **31** | 10 | 11 | 10 | 0,9 | 0,9 | **25%** | 2,00 | **0%** | 1,00 | 1,00 | 1,25 | 2,00 | 6,82 |

**FIGURE 11.2** Wards sheet: risk factors and MAPO index.

| WARD CODE | WARD - UNIT (MANDATORY FIELD) | MAPO Index | Additional Notes | Nr OF INJURIES FOR MPH | INJURIES DATA | | | | |
|---|---|---|---|---|---|---|---|---|---|
| | | | | | INCIDENCE RATE (100 full time workers) | Nr OF WORKING DAYS LOST DUE TO MPH INJURIES | AVERAGE OF WORKING DAYS LOST DUE TO MPH INJURIES | Nr OF INJURIES WITH ACUTE LUMBAGO | Nr OF INJURIES AT SHOULDER |
| 15 | MEDICINE 1 | 5,85 | | 5 | 33,33 | 150 | 30,0 | 2 | 2 |
| 1 | SURGERY 1 | 0,69 | | 11 | 78,57 | 77 | 7,0 | 2 | 3 |
| 33 | GERIATRIC | 8,25 | | 8 | 44,44 | 55 | 6,9 | 2 | 5 |
| 2 | CARDIOLOGIC | 2,56 | | 7 | 38,89 | 60 | 8,6 | 3 | 4 |
| 3 | SURGERY 2 | 2,11 | | 3 | 16,67 | 45 | 15,0 | 1 | 2 |
| 32 | MEDICINE 2 | 8,06 | | 10 | 76,92 | 72 | 7,2 | 3 | 7 |
| 11 | GYNECOLOGY | 3,21 | | 2 | 12,50 | 32 | 16,0 | 1 | 2 |
| 21 | ORTHOPEDY | 16,77 | | 20 | 90,91 | 188 | 9,4 | 5 | 8 |
| 13 | GASTROEN-TEROLOGY | 1,64 | | 4 | 25,00 | 26 | 6,5 | 1 | 3 |
| 38 | NEUROLOGY | 6,82 | | 11 | 45,83 | 50 | 4,5 | 2 | 4 |

**FIGURE 11.3** Wards sheet: risk mapping and injuries data.

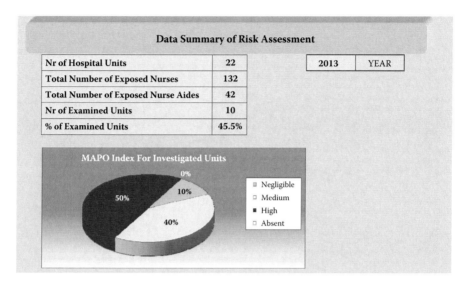

**FIGURE 11.4** Summary and graphs sheet: data summary and MAPO index for investigated units.

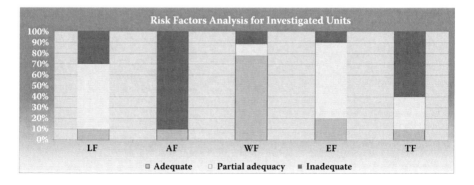

**FIGURE 11.5** Summary and graphs sheet: risk factors analysis for all units analyzed.

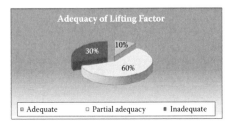

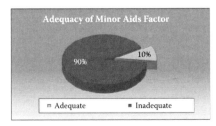

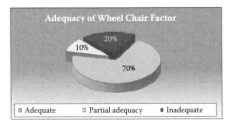

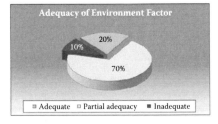

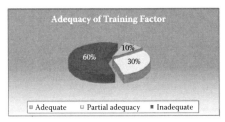

**FIGURE 11.6** Summary and graphs sheet: adequacy of MAPO factors.

| INJURIES DATA | |
|---|---|
| Total Number of Injuries for MPH | 81 |
| Incidence Rate (100 Full-Time Workers) | 46,55 |
| Total Number of Working Days Lost Due to MPH Injuries | 755 |
| Average of Working Days Lost Due to MPH Injuries | 9 |
| Total Number of Injuries with Acute Lumbago | 22 |
| Total Number of Injuries at Shoulder | 40 |

**FIGURE 11.7** Summary and graphs sheet: injuries data.

# 12 Examples of Manual Patient-Handling Risk Assessment in Various Wards

Four examples of MAPO (*movimentazione e assistenza pazienti ospedalizzati*—movement and assistance for hospitalized patients) risk assessments in wards and other departments are shown in this chapter. Each example begins with a brief overview of the main organizational aspects (section A) of the ward, which help to define the main risk factors; next there is the completed data collection sheet (section B), followed by a possible preventive strategy for reducing the level of risk (section C).

The first two examples refer to general and surgical wards, and include detailed descriptions of routine patient-handling tasks, the ward environment, and the equipment on hand, as if the reader were participating in the on-site inspection. This approach ensures that the precompiled MAPO checklist is correctly interpreted (section B).

The other two examples refer to a surgical unit and a radiology department, respectively.

The final section of each example refers to potential corrective actions for reducing the specific risk, with a simulation of any changes to the risk index in the event that these preventive strategies are adopted.

The following examples are provided:

Description of medical ward (Figure 12.1)
Completed form for medical ward (Figure 12.2)
Choice of aids for risk reduction in medical ward (Figure 12.3)

Description of surgical ward (Figure 12.4)
Completed form for surgical ward (Figure 12.5)
Choice of aids for risk reduction in surgical ward (Figure 12.6)

Description of operating rooms (Figure 12.7)
Completed form for operating rooms (Figure 12.8)
Choice of aids for risk reduction in operating rooms (Figure 12.9)

Description of radiology department (Figure 12.10)
Completed form for radiology department (Figure 12.11)
Choice of aids for risk reduction in radiology department (Figure 12.12)

| No. beds: 28 | Patient care staff: 15 | No. operators over three shifts: nine (four morning, three afternoon, two night) |
|---|---|---|

## Brief Overview of Ward

- Average duration of stay is 10 days.
- High number of noncooperative (NC) patients require total lifting (NC = 12).
- Low number of staff are engaged in manual patient handling over three shifts (Op = nine).
- Partially cooperative (PC = 10) patients are primarily *elderly with multiple pathologies.*
- The ward has a sliding sheet, a sliding board, and a height-adjustable stretcher.
- The environment features a lack of space in the bathrooms.
- There are nine wheelchairs totally lacking in ergonomic requirements and two with nonremovable armrests.
- All beds are nonadjustable and only one is a two-segment model.
- Nine of the current staff members (less than 75%) received training 2.5 years earlier, in the form of a 6-hour theoretical/practical course.
- The effectiveness of the training was not measured.

## Brief Overview of Patient-Handling Activities

The most frequent manual patient-handling activities are

- Transfer PC patients from bed to wheelchair and vice versa: twice in the morning to take patients to diagnostic ward and return them to bed, and twice in the afternoon.
- NC patients remain in bed.
- Pull NC and PC patients up in bed approximately twice in the morning, twice in the afternoon, and once during the night shift. Turn NC patients over in bed to change posture (twice per shift).
- Transfer NC patients from bed to stretcher for diagnostic exams (the only aided transfers performed): both from bed to stretcher and from stretcher to bed (two tasks in the morning).

Percentage of aided **total** patient lifts = 29%
Percentage of aided **partial** patient lifts = 0%

## Remarks and Organizational Aspects

The risk level is high due to the inadequacy of the lifting aids and to the large number of disabled patients versus the number of operators present. Moreover, staff training, ward beds, and lifting aids are all inadequate. Frequent patient-handling activities under such conditions call for corrective action to be taken as a top priority in order to reduce risk levels.

**FIGURE 12.1**   Description of medical ward.

## DATA COLLECTION SHEET – RISK ASSESSMENT FOR MANUAL PATIENT HANDLING IN WARDS

### 1. INTERVIEW

| DESCRIPTION OF THE HEALTHCARE FACILITY | | |
|---|---|---|
| HOSPITAL : example 1B | WARD : MEDICINE | WARD CODE: |
| Nr BEDS : 28 | AVERAGE HOSPITAL STAY (days) : 10 | DATE : |

**Nr OF OPERATORS ENGAGED IN MPH:** indicate the total number of operators per job category

| Nursing staff: 4 | Nurses aides: 11 | Other: |
|---|---|---|

**Nr of OPERATORS ENGAGED IN MPH OVER 3 SHIFTS:** indicate the number of operators on duty per shift

| SHIFT | Morning | Afternoon | night |
|---|---|---|---|
| Shift schedule: (00:00 to 00:00) | From 7:00 am to 2:00 pm | From 2:00 pm to 9:00 pm | From 9:00 pm to 7:00 am |
| Nr of operators over entire shift | 4 | 3 | 2 |

| (A) Total operators over entire shift = | 9 |
|---|---|

**Nr of OF PART-TIME OPERATORS:** indicate the exact number of hours worked and calculate them as unit fractions (in relation to the overall duration of the shift).

| Nr of part-time operators present | Hours worked in shift: (00:00 to 00:00) | Unit fraction | (unit fraction by Nr of operators) |
|---|---|---|---|
| | from_____to_____ | | |
| | from_____to_____ | | |
| | from_____to_____ | | |

| (B) Total operators (as unit fractions) present by shift duration = | 0 |
|---|---|

| TOTAL Nr OF OPERATORS ENGAGED IN MPH OVER 24 HOURS (Op): add the total number of operators present over the entire shift (A) to the total number of part-time operators (B) | 9 | Op |
|---|---|---|

Is the work carried out by two nurses? If it is, indicate the number of 2-nurse teams per shift:
1° morning ____2_____    2° afternoon____1_____    3° night ____1_____

### TYPE OF PATIENTS:

"Totally Non-Cooperative" patients (**NC**) are patients who need to be fully lifted in transfer/repositioning operations. "Partially Cooperative" patients (**PC**) are patients who need only partial lifting.

DISABLED PATIENTS (**D**)_____22_____ (indicate average number per day)
Non-Cooperative patients (**NC**) Nr _____12_____    Partially Cooperative patients (**PC**) Nr ___10_____

| DISABLED PATIENTS | Nr NC | Nr PC |
|---|---|---|
| Elderly with multiple concomitant diseases | 12 | 10 |
| Hemiplegic | | |
| Surgical | | |
| Severe stroke | | |
| Dementia | | |
| Other neurologic diseases | | |
| Fracture | | |
| Bariatric | | |
| Other | | |
| Total | 12 | 10 |

**FIGURE 12.2**    Completed form for medical ward. *(continued)*

| OPERATOR EDUCATION AND TRAINING | | | | | |
|---|---|---|---|---|---|
| **EDUCATION AND TRAINING** | | | **INFORMATION** | | |
| Attended theoretical/practical course | ■ YES | ☐ NO | Training only on how to use equipment | ☐ YES | ☐ NO |
| if YES, how many months ago? and how many hours/operator | Months __30__ hours __6__ | | Only provided brochures on MPH | ☐ YES | ☐ NO |
| if YES, how many operators? | 9 | | if YES, how many operators? | | |
| Was EFFECTIVENESS measured and documented in writing? | | | ☐ YES | | ■ NO |

| PATIENT HANDLING TASKS CURRENTLY CARRIED OUT IN ONE SHIFT: | | | | | | |
|---|---|---|---|---|---|---|
| MANUAL HANDLING: describe routine tasks involving total or partial patient lifting | Total lifting (TL) WITHOUT EQUIPMENT | | | Partial Lifting (PL) WITHOUT EQUIPMENT | | |
| indicate the **number of tasks** per shift involving manual patient handling | morning | afternoon | night | morning | afternoon | night |
| | A | B | C | D | E | F |
| ☐ pulling up in bed | ■□□ | ■□□ | ■□□ | ■□ | ■□□ | ■□□ |
| ☐ turning over in bed (to change position) | | | | ■□□ | ■□□ | ■□□ |
| ☐ bed-to-wheelchair and vice versa | □□ | □□ | □□ | ■□ | ■□ | □□ |
| ☐ lifting from seated to upright position | | | | □□ | □□ | □□ |
| ☐ bed-to-stretcher and vice versa | □□ | □□ | □□ | □□ | □□ | □□ |
| ☐ wheelchair-to-toilet and vice versa | □□ | □□ | □□ | □□ | □□ | □□ |
| ☐ other | □□ | □□ | □□ | □□ | □□ | □□ |
| ☐ other | □□ | □□ | □□ | □□ | □□ | □□ |
| TOTAL: calculate the total for each column | 2 | 2 | 1 | 6 | 6 | 3 |
| **Number of total (TL) or partial (PL) manual lifting tasks** | A+B+C = TL | 5 | | D+E+F=PL | 15 | |

| AIDED HANDLING: describe routine tasks involving total or partial patients lifting using available equipment | Total lifting (TL) AIDED | | | Partial Lifting (PL) AIDED | | |
|---|---|---|---|---|---|---|
| Indicate the number of tasks per shift involving aided patient handling | morning | afternoon | night | morning | afternoon | night |
| | G | H | I | L | M | N |
| ☐ pulling up in bed | □□□□ | □□□□ | □□□□ | □□□□ | □□□□ | □□□□ |
| ☐ turning over in bed (to change position) | | | | □□□□□ | □□□□□ | □□□□□ |
| ☐ bed-to-wheelchair and vice versa | □□ | □□ | □□ | □□ | □□ | □□ |
| ☐ lifting from seated to upright position | | | | □□ | □□ | □□ |
| ☐ bed-to-stretcher and vice versa | ■■ | □□ | □□ | □□ | □□ | □□ |
| ☐ wheelchair-to-toilet and vice versa | □□ | □□ | □□ | □□ | □□ | □□ |
| ☐ other | □□ | □□ | □□ | □□ | □□ | □□ |
| ☐ other | □□ | □□ | □□ | □□ | □□ | □□ |
| TOTAL: calculate the total for each column | 2 | | | | | |
| **AIDED handling total (ATL) or partial (APL) lifting** | G+H+I = ATL | 2 | | L+M+N=APL | 0 | |
| **% OF AIDED TOTAL LIFTING OPERATIONS (% ATL)** | ATL / (TL + ATL) | 2/7 =29% | | | | |
| **% OF AIDED PARTIAL LIFTING OPERATIONS (% APL)** | | | | APL / (PL + APL) | 0% | |

**FIGURE 12.2** *(continued)*    Completed form for medical ward. *(continued)*

FONDAZIONE IRCCS CA' GRANDA
Ospedale Maggiore Policlinico
CLINICA DEL LAVORO – MILAN (ITALY)
ERGONOMICS SECTION

## 2.ON SITE INSPECTION

EQUIPMENT FOR DISABLED PATIENT LIFTING/TRANSFER *

| EQUIPMENT DESCRIPTION | | Nr | Lack of essential requirements | | Lack of adaptability to patients or environment | | Lack of maintenance | |
|---|---|---|---|---|---|---|---|---|
| LIFTING EQUIPMENT type : | | | YES | NO | YES | NO | YES | NO |
| LIFTING EQUIPMENT type : | | | YES | NO | YES | NO | YES | NO |
| LIFTING EQUIPMENT type : | | | YES | NO | YES | NO | YES | NO |
| Adjustable STRETCHER type : | | 1 | YES | NO | YES | NO | YES | NO |
| Adjustable STRETCHER type : | | | YES | NO | YES | NO | YES | NO |

OTHER AIDS (MINOR AIDS):

| EQUIPMENT DESCRIPTION | | Nr | Lack of essential requirements | | Lack of adaptability to patients or environment | | Lack of maintenance | |
|---|---|---|---|---|---|---|---|---|
| SLIDING SHEETS | | 1 | YES | NO | YES NO | | YES NO | |
| STANDING HOISTS type: | | | YES | NO | YES | NO | YES | NO |
| ERGONOMIC BELTS: | | | YES | NO | YES | NO | YES | NO |
| SLIDING BOARDS: | | 1 | YES | NO | YES NO | | YES NO | |
| OTHER: | | | YES | NO | YES | NO | YES | NO |

* N.B.: Attach floor plan to assess available space for more equipment and if there is an equipment storage room

| WHEELCHAIRS: | Score | Type of wheelchair | | | | | | | |
|---|---|---|---|---|---|---|---|---|---|
| WHEELCHAIR FEATURES AND INADEQUACY SCORE | | A Nr | B Nr | C Nr | D Nr | E Nr | F Nr | Total Nr of wheelchairs |
| | | 9 | 2 | | | | | |_11__| |
| Poor maintenance | | | | | | | | |
| Malfunctioning brakes | 1 | x | | | | | | |
| Non-removable armrest | 1 | x | x | | | | | |
| Non-removable footrest | | | | | | | | |
| Cumbersome backrest | 1 | x | | | | | | Total wheelchair score: |
| Width exceeding 70 cm | 1 | Cm (72) x | Cm | Cm | Cm | Cm | Cm | |
| **Column score** (Nr of wheelchairs x sum of scores) | | 36 | 2 | | | | | 38 |

**MEAN WHEELCHAIRS SCORE (MSWh) = Total wheelchair score / Nr of wheelchairs = |_3,4_| MSWh**

**FIGURE 12.2** *(continued)*    Completed form for medical ward. *(continued)*

FONDAZIONE IRCCS CA' GRANDA
OSPEDALE MAGGIORE POLICLINICO
DI.P.RO. DEL LAVORO – MILANO ITALY
ERGONOMICS SECTION

**STRUCTURAL FEATURES OF ENVIRONMENT BATHROOMS** (centralized or individual in rooms)

**TYPES OF BATHROOMS WITH SHOWER/BATH:**

| BATHROOMS WITH SHOWER/BATH: FEATURES AND INADEQUACY SCORE | Score | TYPE OF BATHROOM WITH SHOWER/BATH | | | | | | | Total Nr of bathrooms |_15_| |
|---|---|---|---|---|---|---|---|---|---|
| | | En-suite | | | Centralized bathrooms | | | | |
| | | Nr 2 | Nr 2 | Nr 10 | Nr 1 | Nr | Nr | Nr | |
| Free space inadequate for use of aids | 2 | X | | X | X | | | | |
| Door opening inwards (not outwards) | | X | X | | | | | | |
| No shower | | | | | | | | | |
| No bath | | X | X | X | | | | | |
| Door width less than 85 cm | 1 | cm | cm | cm | Cm | cm | cm | cm | Total bathroom score |
| Non-removable obstacles | 1 | | | | | | | | |
| **Column score** (Nr bathrooms x sum of scores) | | 4 | 0 | 20 | 2 | | | | 26 |

Mean bathroom score (**MBS**) = Total bathroom score/total Nr bathrooms :     |_1,73_| MBS

**TOILETS (WC):**

| TOILETS: FEATURES AND INADEQUACY SCORE | Score | TYPE OF TOILETS (WC) | | | | | | | Total Nr of toilets (WC) |_15_| |
|---|---|---|---|---|---|---|---|---|---|
| | | En-suite | En-suite | En-suite | Centralized bathrooms | | | | |
| | | Nr 14 | Nr | Nr | Nr 1 | Nr | Nr | Nr | |
| Free space insufficient to turn around wheelchair | 2 | X | | | X | | | | |
| Door opening inwards (not outwards) | | | | | | | | | |
| Insufficient height of WC (below 50 cm) | 1 | | | | | | | | |
| WC without grab bars* | 1 | X | | | X | | | | |
| Door width less than 85 cm | 1 | | | | | | | | Total WC score: |
| Space at side of WC less than 80 cm | 1 | X | | | X | | | | |
| **Column score** (Nr toilets x sum of scores) | | 56 | | | 4 | | | | 60 |

* if GRAB BARS are present but inadequate, indicate reason for inadequacy in notes and count as absent

Mean WC score (**MSWC**) = total WC score/Nr WCs: |___4___| MSWC

NOTES

_____

_____

_____

**FIGURE 12.2** *(continued)*    Completed form for medical ward. *(continued)*

| PATIENT ROOM CONFIGURATION | | PATIENT ROOMS | | | | | |
|---|---|---|---|---|---|---|---|
| ROOMS: FEATURES AND INADEQUACY SCORE | Score | Nr of rooms **14** | Nr of rooms | Nr of rooms | Nr of rooms | Nr of rooms | |
| Number of beds per room | | **2** | | | | | |
| Space between beds or between bed and wall less than 90 cm | 2 | | | | | | Total Nr of rooms |
| Space between foot of bed and wall less than 120 cm | 2 | | | | | | \|_14_\| |
| Presence of non-removable obstacles | | | | | | | |
| Fixed beds with height less than 70 cm | | Cm Nr | Cm Nr | Cm Nr | Cm Nr | Cm Nr | |
| Unsuitable bed that needs to be partially lifted | 1 | x | | | | | |
| Inadequate side flaps | | | | | | | |
| Door width | | Cm | cm | cm | cm | cm | |
| Space between bed and floor less than 15 cm | 2 | cm | cm | cm | cm | cm | |
| Beds with 2 wheels or no wheels | | | | | | | Total room |
| Height of armchair seat less than 50 cm | 0,5 | | | | | | score: |
| **Column score (Nr of rooms x sum of scores)** | | **14** | | | | | **14** |

Mean room score (**MSR**) = total ward score /total Nr rooms    \|__1___\| **MSR**

INDICATE IF BATHROOMS (OR WHEELCHAIRS) ARE NOT USED BY DISABLED PATIENTS (CONFINED TO BED)
☐ YES   ☐ NO

**MEAN ENVIRONMENT SCORE: MSB + MSWC + MSR =   \|__6,73___\| MSENV  (1,73+4+1)**

| HEIGHT-ADJUSTABLE BEDS | | | | | | |
|---|---|---|---|---|---|---|
| DESCRIPTION OF BEDS | | Nr | Electric adjustable | Mechanical adjustable | Nr of sections | Manual lifting of bed head or foot |
| BED A: |  | | YES   NO | YES   NO | 1  2  3  4 | YES   NO |
| BED B: | | | YES   NO | YES   NO | 1  2  3  4 | YES   NO |
| BED C: | | | YES   NO | YES   NO | 1  2  3  4 | YES   NO |
| BED D: | | | YES   NO | YES   NO | 1  2  3  4 | YES   NO |

**FIGURE 12.2** *(continued)*    Completed form for medical ward. *(continued)*

FONDAZIONE IRCCS CA' GRANDA
OSPEDALE MAGGIORE POLICLINICO
CLINICA DEL LAVORO – MILAN (ITALY)
EPM – EX ICP – REGIONE LOMBARDIA

**MAPO WARD SUMMARY**          Date _____

| Hospital: EXAMPLE 1B | Ward: MEDICINE | Ward code: |
|---|---|---|

Nr of Beds ___28_____          Nr of Operators (**Op**)  |___9___|

Nr of non-cooperative patients **NC** ___12_____     Nr of partially cooperative patients **PC** ___10___

| LIFTING DEVICES FACTOR (LF) | VALUE OF LF | |
|---|---|---|
| LIFTING AIDS ABSENT OR PRESENT BUT NEVER USED | 4 | |
| ABSENT OR INADEQUATE (% ATL ≤ 90%) + INSUFFICIENT Lifting Devices | 4 | |
| INSUFFICIENT OR INADEQUATE Lifting Devices | 2 | |__4__| LF |
| ADEQUATE AND SUFFICIENT Lifting Devices | 0.5 | |

| MINOR AIDS FACTOR (AF) | VALUE OF AF | |
|---|---|---|
| Minor Aids ABSENT OR INSUFFICIENT | 1 | |
| Minor Aids SUFFICIENT AND ADEQUATE (% APL ≥ 90%) | 0.5 | |__1__| AF |

**WHEELCHAIR FACTOR (WF)**

| Mean wheelchair score (MSWh) | 0 – 1.33 | | 1.34 – 2.66 | | 2.67 - 4 | | |
|---|---|---|---|---|---|---|---|
| Numerical sufficiency | NO | YES | NO | YES | NO | YES | |__1,5__| WF |
| VALUE OF WF | 1 | 0.75 | 1.5 | 1.12 | 2 | 1.5 | |

**ENVIRONMENT FACTOR**

| Mean environment score (MSENV) | 0 – 5.8 | 5.9 – 11.6 | 11.7 – 17.5 | |
|---|---|---|---|---|
| VALUE OF EF | 0.75 | 1.25 | 1.5 | |__1,25__| EF |

| TRAINING FACTOR | VALUE OF TF | |
|---|---|---|
| Adequate training | 0.75 | |
| Only information | 1 | |
| No training | 2 | |__2__| TF |

**MAPO INDEX**

$$\text{MAPO INDEX} = (|\_1,33\_| \times |\_4\_| + |\_1,1\_| \times |\_1\_|) \times |\_1,5\_| \times |\_1,25\_| \times |\_2\_| = \mathbf{24,17}$$

NC/OP   LF   PC/OP   AF   WF   EF   TF

| MAPO INDEX |
|---|
| 0 |
| 0.1 – 1.5 |
| 1.51 – 5 |
| > 5 |

**FIGURE 12.2** *(continued)* Completed form for medical ward.

FONDAZIONE IRCCS CA' GRANDA
OSPEDALE MAGGIORE POLICLINICO
CLINICA DEL LAVORO – MILAN ITALY
ERGONOMIC SECTION

**PATIENT HANDLING RISK IN <u>WARDS</u>: CORRECTIVE ACTIONS INVOLVING LIFTING EQUIPMENT**

HOSPITAL : ____ example 1 C _____       WARD :_____MEDICINE_____

## DATA OBTAINED FROM THE MAPO FORM:

| MAPO INDEX | EXPOSURE LEVEL |
|---|---|
| 0 | ABSENT |
| 0.1 – 1.5 | NEGLIGIBLE |
| 1.51 – 5 | MEDIUM |
| > 5 | HIGH |

PERCENTAGE OF AIDED TOTAL LIFTS   (ATL/MTL+ ATL)   29 %

PERCENTAGE OF AIDED PARTIAL LIFTS   (APL/MPL+APL)   0 %

MAPO   $= ( |\_1.33\_| \times |\_4\_| + |\_1.1\_| \times |1| ) \times |1.5| \times |1.25| \times |\_2\_| =$ **24,17**
INDEX        NC/OP        LF        PC/OP        AF        WF        EF        TF

<u>MANUAL DISABLED PATIENT HANDLING OPERATIONS REQUIRING THE USE OF LIFTING AIDS</u>

| MANUAL PATIENT HANDLING: describe routine tasks involving total or partial patient lifting | Total lifting (TL) WITHOUT EQUIPMENT | | | Partial lifting (PL) WITHOUT EQUIPMENT | | |
|---|---|---|---|---|---|---|
| Indicate the **number of tasks** per shift involving manual patient handling | morning | afternoon | night | morning | afternoon | night |
| | A | B | C | D | E | F |
| □ pulling up in bed | | | | | | |
| □ turning over in bed (to change position) | | | | | | |
| □ bed-to-wheelchair and vice versa | | | | | | |
| □ lifting from seated to upright position | | | | | | |
| □ bed-to-stretcher and vice versa | | | | | | |
| □ wheelchair-to-toilet and vice versa | | | | | | |
| □ other | | | | | | |
| □ other | | | | | | |
| TOTAL: calculate the total for each column | 2 | 2 | 1 | 6 | 6 | 3 |
| Number of total (MTL) or partial (MPL) manual lifting tasks | A+B+C = MTL | | 5 | D+E+F=MPL | | 15 |

**PROPOSED SHORT-TERM CORRECTIVE ACTIONS**

— provide 2 sets of sliding sheets (a set consists of two long sliding sheets – one tubular sheet – one sheet for obese patients that slides in two directions), one set per pair of operators;

— replace beds with ergonomic beds, all electrically operated, with 3 or 4 segments;

— provide ergonomic belts (in several sizes).

— supply all staff with specific training on how to use lifting equipment;

☐ ERGONOMIC BEDS (N°= |28| )

☐ SLIDING SHEETS (N°= |_|_2_| ) **SET**    Size _____    Type _____

☐ ERGONOMIC BELTS  (N°= |_|_4_| )    ☐ TROLLEYS (N°= |_|_| )

**FIGURE 12.3**   Choice of aids for risk reduction in medical ward. *(continued)*

FONDAZIONE IRCCS CA' GRANDA
OSPEDALE MAGGIORE POLICLINICO
CLINICA DEL LAVORO – MILAN (ITALY)
ERGONOMICS SECTION

## PREDICTED FINAL MAPO INDEX FOLLOWING "SHORT TERM CORRECTIVE ACTION"

DISABLED PATIENT HANDLING OPERATIONS REMAINING MANUAL:

| MANUAL PATIENT HANDLING: routine tasks involving total or partial patient lifting | Total lifting (TL) WITHOUT EQUIPMENT | | | Partial Lifting (PL) WITHOUT EQUIPMENT | | |
|---|---|---|---|---|---|---|
| | morning | afternoon | night | morning | afternoon | night |
| | A | B | C | D | E | F |
| ☐ pulling up in bed | ☐☐☐☐ | ☐☐☐☐ | ☐☐☐☐ | ☐☐☐☐ | ☐☐☐☐ | ☐☐☐☐ |
| ☐ turning over in bed (to change position) | | | | ☐☐☐☐☐ | ☐☐☐☐☐ | ☐☐☐☐☐ |
| ☐ bed-to-wheelchair and vice versa | ☐☐ | ☐☐ | ☐☐ | ■ | ☐☐ | ☐☐ |
| ☐ lifting from seated to upright position | | | | ☐☐ | ☐☐ | ☐☐ |
| ☐ bed-to-stretcher and vice versa | ☐☐ | ☐☐ | ☐☐ | ☐☐ | ☐☐ | ☐☐ |
| ☐ wheelchair-to-toilet and vice versa | ☐☐ | ☐☐ | ☐☐ | ☐☐ | ☐☐ | ☐☐ |
| ☐ other | ☐☐ | ☐☐ | ☐☐ | ☐☐ | ☐☐ | ☐☐ |
| ☐ other | ☐☐ | ☐☐ | ☐☐ | ☐☐ | ☐☐ | ☐☐ |
| TOTAL: calculate the total for each column | | | | 2 | 2 | |

OVERALL PATIENT HANDLING OPERATIONS AIDED WITH LIFTING EQUIPMENT

| AIDED HANDLING: routine tasks involving total or partial patient lifting using available equipment | Total lifting (TL) AIDED | | | Partial Lifting (PL) AIDED | | |
|---|---|---|---|---|---|---|
| | morning | afternoon | night | morning | afternoon | night |
| | G | H | I | L | M | N |
| ☐ pulling up in bed | ■☐☐ | ■☐☐☐ | ☐☐☐☐ | ■☐☐ | ☐☐☐☐ | ☐☐☐☐ |
| ☐ turning over in bed (to change position) | | | | ■☐ | ☐☐☐☐ | ■☐☐ |
| ☐ bed-to-wheelchair and vice versa | ☐☐ | ☐☐ | ☐☐ | ☐☐ | ☐☐ | ☐☐ |
| ☐ lifting from seated to upright position | | | | ☐☐ | ☐☐ | ☐☐ |
| ☐ bed-to-stretcher and vice versa | ■☐ | ☐☐ | ☐☐ | ☐☐ | ☐☐ | ☐☐ |
| ☐ wheelchair-to-toilet and vice versa | ☐☐ | ☐☐ | ☐☐ | ☐☐ | ☐☐ | ☐☐ |
| ☐ other | ☐☐ | ☐☐ | ☐☐ | ☐☐ | ☐☐ | ☐☐ |
| ☐ other | ☐☐ | ☐☐ | ☐☐ | ☐☐ | ☐☐ | ☐☐ |
| TOTAL: calculate the total for each column | 4 | 2 | 1 | 4 | 4 | 3 |

### NEW PERCENTAGE OF AIDED LIFTING OPERATIONS

PERCENTAGE OF AIDED TOTAL LIFTING OPERATIONS (7/7) |100%|

PERCENTAGE OF AIDED PARTIAL LIFTING OPERATIONS (11/15) |73%|

**PREDICTED MAPO INDEX FOLLOWING ADOPTION OF LIFTING AIDS AND STAFF TRAINING**

$$\text{MAPO} = (|\_1.33\_| \times |\_0.5\_| + |\_1.1\_| \times |\_1\_|) \times |\_1.5| \times |\_1.25| \times |\_0.75\_| = \mathbf{2.5}$$

INDEX     NC/OP    LF     PC/OP    AF      WF      EF      TF

**PROPOSED LONG TERM CORRECTIVE ACTION WITH IMPROVED QUALITY OF CARE**

The environment still remains to be improved, and the percentage of aided lifting operations must increase substantially, in light of the extremely poor ratio of disabled patients to number of operators.

Therefore, over the long term, it will be necessary for a standing hoist to be used for partially cooperative patients and for the number of staff members to be increased (two more operators over the 24 hours).

**PREDICTED MAPO INDEX FOLLOWING LONG TERM CORRECTIVE ACTIONS**

$$\text{MAPO} = (|\_1.1\_| \times |\_0.5\_| + |\_0.9\_| \times |\_0.5\_|) \times |\_1.5| \times |\_1.25| \times |\_0.75\_| = \mathbf{1.41}$$

INDEX     NC/OP    LF     PC/OP    AF      WF      EF      TF

**FIGURE 12.3** *(continued)*    Choice of aids for risk reduction in medical ward.

| No. beds: 25 | Patient care staff: 19 | No. operators over three shifts: 11 (five morning, four afternoon, two night) |
|---|---|---|

### Brief Overview of Ward

- Average duration of stay is 6 days.
- The patients who need lifting are those going into surgery and on day 1 postsurgery. Three surgeries are performed per day, making for a total of six NC patients.
- In addition, there are ten PC patients requiring partial lifting.
- Equipment: there is only one sliding sheet (never used) and all beds are height adjustable but not ergonomic (two segments, mechanically powered, upper segment to be lifted).
- Overall, the environment has few shortcomings.
- There are six wheelchairs, somewhat lacking in ergonomic requirements.
- Training is partly adequate since 53% of staff (10 out of 19) received training over the previous year.

### Brief Overview of Patient-Handling Activities

The most frequent manual patient-handling activities are

- Pull NC and PC patients up in bed and turn over to change posture.
- Transfer patients from bed to wheelchair and vice versa and from wheelchair to toilet and vice versa (only partial lifting) in the morning and afternoon shifts. These tasks are performed only once during the first two shifts, but are counted as two (going and coming back).
- Lift from bed to stretcher and vice versa after surgery and for diagnostic tests during the first two shifts.

> Percentage of aided partial patient lifts = 0%
> Percentage of aided **total** patient lifts = 0%

### Remarks and Organizational Aspects

The greatest risk is due to the almost total absence of lifting equipment (except for one sliding sheet, never used) and only partially adequate staff training. The organizational aspects (i.e., number of operators, grouped into pairs per bed) could be readily improved. The inadequacy of the ward beds (only two segments) needs to be emphasized, as well as of the lifting equipment and standard of care.

**FIGURE 12.4** Description of surgical ward.

FONDAZIONE IRCCS CA' GRANDA
OSPEDALE MAGGIORE POLICLINICO
CLINICA DEL LAVORO – MILAN (ITALY)
ERGONOMICS SECTION

**DATA COLLECTION SHEET – RISK ASSESSMENT FOR MANUAL PATIENT HANDLING IN WARDS**

### 1. INTERVIEW

| DESCRIPTION OF THE HEALTHCARE FACILITY | | |
|---|---|---|
| HOSPITAL : example 2C | WARD : SURGERY | WARD CODE: |
| Nr BEDS : 25 | AVERAGE HOSPITAL STAY (days) : 6 | DATE : |

| Nr OF OPERATORS ENGAGED IN MPH: indicate the total number of operators per job category | | |
|---|---|---|
| Nursing staff: 9 | Nurses aides: 10 | Other: |

Nr of OPERATORS ENGAGED IN MPH OVER 3 SHIFTS: indicate the number of operators on duty per shift

| SHIFT | Morning | Afternoon | night |
|---|---|---|---|
| Shift schedule: (00:00 to 00:00) | From 7:00 am to 2:00 pm | From 2:00 pm to 9:00 pm | From 9:00 pm to 7:00 am |
| Nr of operators over entire shift | 5 | 4 | 2 |

| (A) Total operators over entire shift = | **11** |
|---|---|

Nr of OF PART-TIME OPERATORS: indicate the exact number of hours worked and calculate them as unit fractions (in relation to the overall duration of the shift).

| Nr of part-time operators present | Hours worked in shift: (00:00 to 00:00) | Unit fraction | (unit fraction by Nr of operators) |
|---|---|---|---|
| | from_____to_____ | | |
| | from_____to_____ | | |
| | from_____to_____ | | |

| (B) Total operators (as unit fractions) present by shift duration = | **0** |
|---|---|

| TOTAL Nr OF OPERATORS ENGAGED IN MPH OVER 24 HOURS (Op): add the total number of operators present over the entire shift (A) to the total number of part-time operators (B) | **11** | **Op** |
|---|---|---|

Is the work carried out by two nurses? If it is, indicate the number of 2-nurse teams per shift:
1° morning ____2_____    2° afternoon____2_____    3° night ____1____

### TYPE OF PATIENTS:

"Totally Non-Cooperative" patients (**NC**) are patients who need to be fully lifted in transfer/repositioning operations. "Partially Cooperative" patients (**PC**) are patients who need only partial lifting.

| DISABLED PATIENTS (**D**)_____16_____ (indicate average number per day) |
|---|
| Non-Cooperative patients (**NC**) Nr _____6_____    Partially Cooperative patients (**PC**) Nr ___10_____ |

| DISABLED PATIENTS | Nr NC | Nr PC |
|---|---|---|
| Elderly with multiple concomitant diseases | | |
| Hemiplegic | | |
| Surgical | 6 | 10 |
| Severe stroke | | |
| Dementia | | |
| Other neurologic diseases | | |
| Fracture | | |
| Bariatric | | |
| Other | | |
| Total | 6 | 10 |

**FIGURE 12.5**    Completed form for surgical ward. *(continued)*

| OPERATOR EDUCATION AND TRAINING | | | | | | | |
|---|---|---|---|---|---|---|---|
| **EDUCATION AND TRAINING** | | | | **INFORMATION** | | | |
| Attended theoretical/practical course | ■YES | ☐NO | | Training only on how to use equipment | | ☐YES | ☐NO |
| if YES, how many months ago? and how many hours/operator | Months __10__ hours ___8___ | | | Only provided brochures on MPH | | ☐YES | ☐NO |
| if YES, how many operators? | **10** | | | if YES, how many operators? | | | |
| Was EFFECTIVENESS measured and documented in writing? | | | | | ☐YES | | ■NO |

| PATIENT HANDLING TASKS CURRENTLY CARRIED OUT IN ONE SHIFT: | | | | | | |
|---|---|---|---|---|---|---|
| **MANUAL HANDLING:** describe routine tasks involving total or partial patient lifting | Total lifting (TL) WITHOUT EQUIPMENT | | | Partial Lifting (PL) WITHOUT EQUIPMENT | | |
| indicate the number of tasks per shift involving manual patient handling | morning | afternoon | night | morning | afternoon | night |
|  | A | B | C | D | E | F |
| ☐ pulling up in bed | ■□□□ | ■□□□ | ■□□□ | ■□□ | ■□□□ | ■□□□ |
| ☐ turning over in bed (to change position) |  |  |  | ■□□□□ | ■□□□□ | ■□□□□ |
| ☐ bed-to-wheelchair and vice versa | □□ | □□ | □□ | ■ | ■ | □□ |
| ☐ lifting from seated to upright position |  |  |  | □□ | □□ | □□ |
| ☐ bed-to-stretcher and vice versa | ■ | ■ | □□ | □□ | □□ | □□ |
| ☐ wheelchair-to-toilet and vice versa | □□ | □□ | □□ | ■ | ■ | □□ |
| ☐ other | □□ | □□ | □□ | □□ | □□ | □□ |
| ☐ other | □□ | □□ | □□ | □□ | □□ | □□ |
| TOTAL: calculate the total for each column | **3** | **3** | **1** | **6** | **6** | **2** |
| Number of total (TL) or partial (PL) manual lifting tasks | A+B+C = TL | **7** | | D+E+F=PL | **14** | |

| AIDED HANDLING: describe routine tasks involving total or partial patients lifting using available equipment | Total lifting (TL) AIDED | | | Partial Lifting (PL) AIDED | | |
|---|---|---|---|---|---|---|
| Indicate the number of tasks per shift involving aided patient handling | morning | afternoon | night | morning | afternoon | night |
|  | G | H | I | L | M | N |
| ☐ pulling up in bed | □□□□ | □□□□ | □□□□ | □□□□ | □□□□ | □□□□ |
| ☐ turning over in bed (to change position) |  |  |  | □□□□□ | □□□□□ | □□□□□ |
| ☐ bed-to-wheelchair and vice versa | □□ | □□ | □□ | □□ | □□ | □□ |
| ☐ lifting from seated to upright position |  |  |  | □□ | □□ | □□ |
| ☐ bed-to-stretcher and vice versa | □□ | □□ | □□ | □□ | □□ | □□ |
| ☐ wheelchair-to-toilet and vice versa | □□ | □□ | □□ | □□ | □□ | □□ |
| ☐ other | □□ | □□ | □□ | □□ | □□ | □□ |
| ☐ other | □□ | □□ | □□ | □□ | □□ | □□ |
| TOTAL: calculate the total for each column |  |  |  |  |  |  |
| AIDED handling total (ATL) or partial (APL) lifting | G+H+I = ATL | **0** | | L+M+N=APL | **0** | |
| % OF AIDED TOTAL LIFTING OPERATIONS (% ATL) | $\frac{ATL}{(TL + ATL)}$ | **0%** | | | | |
| % OF AIDED PARTIAL LIFTING OPERATIONS (% APL) | | | | $\frac{APL}{(PL + APL)}$ | **0%** | |

**FIGURE 12.5** *(continued)*   Completed form for surgical ward. *(continued)*

FONDAZIONE IRCCS CA' GRANDA
OSPEDALE MAGGIORE POLICLINICO
CLINICA DEL LAVORO - MILAN (ITALY)
ERGONOMICS SECTION

## 2.ON-SITE INSPECTION

EQUIPMENT FOR DISABLED PATIENT LIFTING/TRANSFER *

| EQUIPMENT DESCRIPTION | | Nr | Lack of essential requirements | | Lack of adaptability to patients or environment | | Lack of maintenance | |
|---|---|---|---|---|---|---|---|---|
| LIFTING EQUIPMENT type : | | | YES | NO | YES | NO | YES | NO |
| LIFTING EQUIPMENT type : | | | YES | NO | YES | NO | YES | NO |
| LIFTING EQUIPMENT type : | | | YES | NO | YES | NO | YES | NO |
| Adjustable STRETCHER type : | | | YES | NO | YES | NO | YES | NO |
| Adjustable STRETCHER type : | | | YES | NO | YES | NO | YES | NO |

OTHER AIDS (MINOR AIDS):

| EQUIPMENT DESCRIPTION | | Nr | Lack of essential requirements | | Lack of adaptability to patients or environment | | Lack of maintenance | |
|---|---|---|---|---|---|---|---|---|
| SLIDING SHEETS | | 1 | YES | **NO** | YES | **NO** | YES | **NO** |
| STANDING HOISTS type: | | | YES | NO | YES | NO | YES | NO |
| ERGONOMIC BELTS: | | | YES | NO | YES | NO | YES | NO |
| SLIDING BOARDS: | | | YES | NO | YES | NO | YES | NO |
| OTHER: | | | YES | NO | YES | NO | YES | NO |

* N.B.: Attach floor plan to assess available space for more equipment and if there is an equipment storage room

| WHEELCHAIRS: | Score | Type of wheelchair | | | | | | Total Nr of wheelchairs |
|---|---|---|---|---|---|---|---|---|
| WHEELCHAIR FEATURES AND INADEQUACY SCORE | | A Nr | B Nr | C Nr | D Nr | E Nr | F Nr | |_6_| |
| | | 3 | 2 | 1 | | | | |
| Poor maintenance | | | | | | | | |
| Malfunctioning brakes | 1 | | X | X | | | | |
| Non-removable armrest | 1 | | X | X | | | | |
| Non-removable footrest | | | | | | | | |
| Cumbersome backrest | 1 | | | X | | | | Total wheelchair score: |
| Width exceeding 70 cm | 1 | Cm | Cm | Cm | Cm | Cm | Cm | |
| Column score (Nr of wheelchairs x sum of scores) | | 0 | 4 | 3 | | | | 7 |

MEAN WHEELCHAIRS SCORE (MSWh) = Total wheelchair score / Nr of wheelchairs = |_1,17_| MSWh

**FIGURE 12.5** *(continued)*   Completed form for surgical ward. *(continued)*

FONDAZIONE IRCCS CA' GRANDA
OSPEDALE MAGGIORE POLICLINICO
CLINICA DEL LAVORO – MILAN (ITALY)
ERGONOMIC SECTION

**STRUCTURAL FEATURES OF ENVIRONMENT BATHROOMS** (centralized or individual in rooms)

**TYPES OF BATHROOMS WITH SHOWER/BATH:**

| BATHROOMS WITH SHOWER/BATH: FEATURES AND INADEQUACY SCORE | Score | TYPE OF BATHROOM WITH SHOWER/BATH | | | | | | | Total Nr of bathrooms |_14_| |
|---|---|---|---|---|---|---|---|---|---|
| | | En-suite | | | Centralized bathrooms | | | | |
| | | Nr **2** | Nr **1** | Nr **10** | Nr **1** | Nr | Nr | Nr | |
| Free space inadequate for use of aids | 2 | x | | x | x | | | | |
| Door opening inwards (not outwards) | | x | x | | | | | | |
| No shower | | | | | | | | | |
| No bath | | x | x | x | | | | | |
| Door width less than 85 cm | 1 | X cm 70 | cm | X cm 70 | Cm | cm | cm | cm | Total bathroom score |
| Non-removable obstacles | 1 | X | | | | | | | |
| **Column score** (Nr bathrooms x sum of scores) | | **8** | **0** | **10** | **0** | | | | **18** |

Mean bathroom score (**MBS**) = Total bathroom score/total Nr bathrooms :    |_1,28_| MBS

**TOILETS (WC):**

| TOILETS: FEATURES AND INADEQUACY SCORE | Score | TYPE OF TOILETS (WC) | | | | | | | Total Nr of toilets (WC) |_14_| |
|---|---|---|---|---|---|---|---|---|---|
| | | En-suite | En-suite | En-suite | Centralized bathrooms | | | | |
| | | Nr **2** | Nr **11** | Nr | Nr **1** | Nr | Nr | Nr | |
| Free space insufficient to turn around wheelchair | 2 | x | | | | | | | |
| Door opening inwards (not outwards) | | x | | | | | | | |
| Insufficient height of WC (below 50 cm) | 1 | | | | | | | | |
| WC without grab bars* | 1 | x | x | | | | | | |
| Door width less than 85 cm | 1 | x | | | | | | | Total WC score: |
| Space at side of WC less than 80 cm | 1 | x | | | | | | | |
| **Column score** (Nr toilets x sum of scores) | | **10** | **11** | | **0** | | | | **21** |

* if GRAB BARS are present but inadequate, indicate reason for inadequacy in notes and count as absent

Mean WC score (**MSWC**) = total WC score/Nr WCs:  |_1,5_| **MSWC**

**NOTES**

_____

_____

_____

**FIGURE 12.5** *(continued)*    Completed form for surgical ward. *(continued)*

| PATIENT ROOM CONFIGURATION | Score | PATIENT ROOMS | | | | | |
|---|---|---|---|---|---|---|---|
| ROOMS: FEATURES AND INADEQUACY SCORE | | Nr of rooms 10 | Nr of rooms 1 | Nr of rooms 2 | Nr of rooms | Nr of rooms | |
| Number of beds per room | | 2 | 1 | 2 | | | Total Nr of rooms |
| Space between beds or between bed and wall less than 90 cm | 2 | | x | | | | |_13_| |
| Space between foot of bed and wall less than 120 cm | 2 | | x | | | | |
| Presence of non-removable obstacles | | | | | | | |
| Fixed beds with height less than 70 cm | | Cm Nr | Cm Nr | Cm Nr | Cm Nr | Cm Nr | |
| Unsuitable bed that needs to be partially lifted | 1 | x | x | x | | | |
| Inadequate side flaps | | | | | | | |
| Door width | | Cm | cm | cm | cm | cm | |
| Space between bed and floor less than 15 cm | 2 | cm | cm | cm | cm | cm | |
| Beds with 2 wheels or no wheels | | | | | | | Total room score: |
| Height of armchair seat less than 50 cm | 0,5 | | | x | | | |
| **Column score (Nr of rooms x sum of scores)** | | 10 | 5 | 3 | | | 18 |

Mean room score (**MSR**) = total ward score /total Nr rooms    |___1,4____| **MSR**

INDICATE IF BATHROOMS (OR WHEELCHAIRS) ARE NOT USED BY DISABLED PATIENTS (CONFINED TO BED)
☐ YES    ■ NO

**MEAN ENVIRONMENT SCORE: MSB + MSWC + MSR =**   |__4,18___| **MSENV**   (1,28 + 1,5 + 1,4)

| HEIGHT-ADJUSTABLE BEDS | | | | | | |
|---|---|---|---|---|---|---|
| DESCRIPTION OF BEDS | | Nr | Electric adjustable | Mechanical adjustable | Nr of sections | Manual lifting of bed head or foot |
| BED A: MECHANICAL | | 25 | YES  **NO** | **YES**  NO | 1 **2** 3 4 | **YES**  NO |
| BED B: | | | YES  NO | YES  NO | 1 2 3 4 | YES  NO |
| BED C: | | | YES  NO | YES  NO | 1 2 3 4 | YES  NO |
| BED D: | | | YES  NO | YES  NO | 1 2 3 4 | YES  NO |

**FIGURE 12.5** *(continued)*    Completed form for surgical ward. *(continued)*

FONDAZIONE IRCCS CA' GRANDA
OSPEDALE MAGGIORE POLICLINICO
CLINICA DEL LAVORO – MILAN (ITALY)
ERGONOMICS SECTION

### MAPO WARD SUMMARY

Date _____

| Hospital: EXAMPLE 2B | Ward: SURGERY | Ward code: |
|---|---|---|

Nr of Beds ___25_____      Nr of Operators (**Op**)   |___11___|

Nr of non-cooperative patients **NC** __6____     Nr of partially cooperative patients **PC** ___10___

| LIFTING DEVICES FACTOR (LF) | VALUE OF LF | |
|---|---|---|
| **LIFTING AIDS ABSENT OR PRESENT BUT NEVER USED** | 4 | |
| **ABSENT OR INADEQUATE (% ATL ≤ 90%) + INSUFFICIENT** Lifting Devices | 4 | |
| **INSUFFICIENT OR INADEQUATE** Lifting Devices | 2 | |___4___| LF |
| **ADEQUATE AND SUFFICIENT** Lifting Devices | 0.5 | |

| MINOR AIDS FACTOR (AF) | VALUE OF AF | |
|---|---|---|
| Minor Aids **ABSENT OR INSUFFICIENT** | 1 | |
| Minor Aids **SUFFICIENT AND ADEQUATE** (% APL ≥ 90%) | 0.5 | |___1___| AF |

**WHEELCHAIR FACTOR (WF)**

| Mean wheelchair score (MSWh) | 0 – 1.33 | | 1.34 – 2.66 | | 2.67 - 4 | | |
|---|---|---|---|---|---|---|---|
| Numerical sufficiency | NO | YES | NO | YES | NO | YES | |___1___| WF |
| VALUE OF WF | 1 | 0.75 | 1.5 | 1.12 | 2 | 1.5 | |

**ENVIRONMENT FACTOR**

| Mean environment score (MSENV) | | 0 – 5.8 | 5.9 – 11.6 | 11.7 – 17.5 | |
|---|---|---|---|---|---|
| VALUE OF EF | | 0.75 | 1.25 | 1.5 | |___0,75___| EF |

**TRAINING FACTOR**

| | VALUE OF TF | |
|---|---|---|
| Adequate training | 0.75 | |
| Only information | 1 | |
| No training | 2 | |___1___| TF |

### MAPO INDEX

MAPO = ( |_0,54_| x |_4_| + |_0,9_| x |_1_| ) x |_1_| x |_0,75_| x |_1_| = **2,32**

INDEX    NC/OP    LF    PC/OP    AF    WF    EF    TF

| MAPO INDEX |
|---|
| 0 |
| 0.1 – 1.5 |
| 1.51 – 5 |
| > 5 |

**FIGURE 12.5** *(continued)*   Completed form for surgical ward.

FONDAZIONE IRCCS CA' GRANDA
OSPEDALE MAGGIORE POLICLINICO
CLINICA DEL LAVORO (MILAN ITALY)
ERGONOMICS SECTION

## PATIENT HANDLING RISK IN <u>WARDS</u>: CORRECTIVE ACTIONS INVOLVING LIFTING EQUIPMENT

HOSPITAL :____ example 2 C _____     WARD :_____SURGERY_____

| MAPO INDEX | EXPOSURE LEVEL |
|---|---|
| 0 | ABSENT |
| 0.1 – 1.5 | NEGLIGIBLE |
| 1.51 – 5 | MEDIUM |
| > 5 | HIGH |

## DATA OBTAINED FROM THE MAPO FORM:

PERCENTAGE OF AIDED TOTAL LIFTS   (ATL/MTL+ ATL)    0 %

PERCENTAGE OF AIDED PARTIAL LIFTS  (APL/MPL+APL)    0 %

MAPO INDEX $= (|\_0{,}54\_| \times |\_4\_| + |\_0{,}9\_| \times |1|) \times |1| \times |0{,}75| \times |\_1\_| =$ **2,32**

         NC/OP      LF      PC/OP    AF     WF     EF       TF

### MANUAL DISABLED PATIENT HANDLING OPERATIONS REQUIRING THE USE OF LIFTING AIDS

| MANUAL PATIENT HANDLING: describe routine tasks involving total or partial patient lifting | Total lifting (TL) WITHOUT EQUIPMENT | | | Partial lifting (PL) WITHOUT EQUIPMENT | | |
|---|---|---|---|---|---|---|
| Indicate the **number of tasks** per shift involving manual patient handling | morning | afternoon | night | morning | afternoon | night |
| | A | B | C | D | E | F |
| □ pulling up in bed | | | | | | |
| □ turning over in bed (to change position) | | | | | | |
| □ bed-to-wheelchair and vice versa | | | | | | |
| □ lifting from seated to upright position | | | | | | |
| □ bed-to-stretcher and vice versa | | | | | | |
| □ wheelchair-to-toilet and vice versa | | | | | | |
| □ other | | | | | | |
| □ other | | | | | | |
| TOTAL: calculate the total for each column | 3 | 3 | 1 | 6 | 6 | 2 |
| Number of total (MTL) or partial (MPL) manual lifting tasks | A+B+C = MTL | 7 | | D+E+F=MPL | 14 | |

### PROPOSED SHORT-TERM CORRECTIVE ACTIONS

— provide 2 sets of sliding sheets (a set consists of two long sliding sheets – one tubular sheet – one sheet for obese patients that slides in two directions), one set per pair of operators;

— provide 1 adjustable stretcher and 1 sliding board;

— supply all staff with specific training on how to use lifting equipment.

| |
|---|
| □ SLIDING SHEETS (N°=\|_\|_2_\|) **SET**      Size _____    Type _____ |
| □ ADJUSTABLE STRETCHER  (N°=\|_\|_1_\|)        □ SLIDING BOARDS (N°=\|_\|_1_\|) |

**FIGURE 12.6** Choice of aids for risk reduction in surgical ward. *(continued)*

### PREDICTED FINAL MAPO INDEX FOLLOWING "SHORT TERM CORRECTIVE ACTION"

<u>DISABLED PATIENT HANDLING OPERATIONS REMAINING MANUAL:</u>

| MANUAL PATIENT HANDLING: routine tasks involving total or partial patient lifting | Total lifting (TL) WITHOUT EQUIPMENT | | | Partial Lifting (PL) WITHOUT EQUIPMENT | | |
|---|---|---|---|---|---|---|
| | morning | afternoon | night | morning | afternoon | night |
| | A | B | C | D | E | F |
| ☐ pulling up in bed | ☐☐☐☐ | ☐☐☐☐ | ☐☐☐☐ | ☐☐☐☐ | ☐☐☐☐ | ☐☐☐☐ |
| ☐ turning over in bed (to change position) | | | | ☐☐☐☐☐ | ☐☐☐☐☐ | ☐☐☐☐☐ |
| ☐ bed-to-wheelchair and vice versa | ☐☐ | ☐☐ | ☐☐ | ▪ | ☐☐ | ☐☐ |
| ☐ lifting from seated to upright position | | | | ☐☐ | ☐☐ | ☐☐ |
| ☐ bed-to-stretcher and vice versa | ☐☐ | ☐☐ | ☐☐ | ☐☐ | ☐☐ | ☐☐ |
| ☐ wheelchair-to-toilet and vice versa | ☐☐ | ☐☐ | ☐☐ | ▪ | ☐☐ | ☐☐ |
| ☐ other | ☐☐ | ☐☐ | ☐☐ | ☐☐ | ☐☐ | ☐☐ |
| ☐ other | ☐☐ | ☐☐ | ☐☐ | ☐☐ | ☐☐ | ☐☐ |
| TOTAL: calculate the total for each column | 0 | 0 | 0 | 4 | 4 | 0 |

<u>OVERALL PATIENT HANDLING OPERATIONS AIDED WITH LIFTING EQUIPMENT</u>

| AIDED HANDLING: routine tasks involving total or partial patient lifting using available equipment | Total lifting (TL) AIDED | | | Partial Lifting (PL) AIDED | | |
|---|---|---|---|---|---|---|
| | morning | afternoon | night | morning | afternoon | night |
| | G | H | I | L | M | N |
| ☐ pulling up in bed | ▪☐☐☐ | ☐☐☐☐ | ☐☐☐☐ | ▪☐☐☐ | ☐☐☐☐ | ☐☐☐☐ |
| ☐ turning over in bed (to change position) | | | | ▪☐☐☐ | ☐☐☐☐ | ☐☐☐☐ |
| ☐ bed-to-wheelchair and vice versa | ☐☐ | ☐☐ | ☐☐ | ☐☐ | ☐☐ | ☐☐ |
| ☐ lifting from seated to upright position | | | | ☐☐ | ☐☐ | ☐☐ |
| ☐ bed-to-stretcher and vice versa | ☐☐ | ☐☐ | ☐☐ | ☐☐ | ☐☐ | ☐☐ |
| ☐ wheelchair-to-toilet and vice versa | ☐☐ | ☐☐ | ☐☐ | ☐☐ | ☐☐ | ☐☐ |
| ☐ other | ☐☐ | ☐☐ | ☐☐ | ☐☐ | ☐☐ | ☐☐ |
| ☐ other | ☐☐ | ☐☐ | ☐☐ | ☐☐ | ☐☐ | ☐☐ |
| TOTAL: calculate the total for each column | 3 | 3 | 1 | 2 | 2 | 2 |

### NEW PERCENTAGE OF AIDED LIFTING OPERATIONS

PERCENTAGE OF AIDED TOTAL LIFTING OPERATIONS (7/7) |100%|

PERCENTAGE OF AIDED PARTIAL LIFTING OPERATIONS (6/14) |43%|

**PREDICTED MAPO INDEX FOLLOWING ADOPTION OF LIFTING AIDS AND STAFF TRAINING**

MAPO $= (|\_0{,}54\_| \times |\_0{,}5\_| + |\_0{,}9\_| \times |1|) \times |1{,}5| \times |\_0{,}75| \times |\_\_0{,}75\_\_|$ = **0,66**
INDEX       NC/OP    LF     PC/OP     AF     WF     EF      TF

**PROPOSED LONG TERM CORRECTIVE ACTION WITH IMPROVED QUALITY OF CARE**

Over the long term it will be necessary for a standing hoist to be used for partially cooperative patients and for ergonomic belts.

**PREDICTED MAPO INDEX FOLLOWING LONG TERM CORRECTIVE ACTIONS**

MAPO $= (|\_0{,}54\_| \times |\_0{,}5\_| + |\_0{,}9\_| \times |\_0{,}5|) \times |1| \times |\_0{,}75| \times |\_0{,}75\_|$ = **0,41**
INDEX       NC/OP    LF     PC/OP     AF     WF     EF      TF

**FIGURE 12.6** *(continued)*  Choice of aids for risk reduction in surgical ward.

| No. procedures/day: 12 | Patient care staff: 18 | No. operators over 24 hours: 15 |

### Brief Overview

- Surgical unit consists of two operating rooms (in use 5 days a week) with staff "on rotation."
- Equipment consists of one sliding board per operating room.
- On average, eight procedures under general anesthesia (GA) and four under local anesthesia (LA) on patients requiring manual handling.
- No staff training, merely distribution of brochures; effectiveness of training not measured.

### Brief Overview of Patient-Handling Activities

- Lifting equipment can only be used in operating room. Some patients (three GA and four LA) have to be brought to surgery from wards and therefore need to be transferred manually from bed to stretcher and vice versa.
- There are four stretchers, none of which is height adjustable.
- Operating room A has insufficient space for using lifting aids while operating room B has an operating table with side flaps.

  Percentage of aided patient lifts = 63%

  Frequency of patient-handling tasks (F) = 38/15 = 2.5

### Remarks and Organizational Aspects

Risk is determined by the *lack of lifting equipment in the wards that patients are brought in from and the lack of staff training.* The operating table has side flaps, making it impossible to use sliding boards; the quality of care is also therefore inadequate.

---

**FIGURE 12.7**   Description of operating rooms.

FONDAZIONE IRCCS CA' GRANDA
OSPEDALE MAGGIORE POLICLINICO
· CLINICA DEL LAVORO – MILAN (ITALY)
ERGONOMICS SECTION

### DATA COLLECTION SHEET:  RISK ASSESSMENT FOR MANUAL PATIENT HANDLING RISK IN SURGICAL UNIT

HOSPITAL:__EXAMPLE 3B_____    SURGICAL  UNIT :____GENERAL_____DATE: _____

## 1. INTERVIEW

| TOTAL OPERATORS (staff performing manual patient handling tasks): | | |
|---|---|---|
| Nurses/scrub nurses:  10 | Nurses aides: 8 | Other: |
| Surgical Unit daily working hours: 8.00 AM –  4 P.M. | Working days: FROM MONDAY TO FRIDAY | |

| OPERATORS (staff performing manual patient handling tasks) PRESENT OVER 24 HOURS: | | |
|---|---|---|
| Nurses/scrub nurses:  8 | Nurses aides: 7 | Other: |

| Indicate the number of operators (Op) as the sum of operators (in all job categories) involved in patient handling and present over 24 hours | 15 | **Op** |
|---|---|---|

### QUANTIFICATION OF SURGICAL PROCEDURES

Nr. of ORs ___2___   Nr. of procedures/year ____3100_____   Average nr. procedures/day   |_12_|
- Average nr. of procedures/day under general anesthesia (**GA**): ____8_____
- Average nr. of procedures/day under local anesthesia :      ____4_____
    Average nr. of procedures/day **not requiring** total/partial patient lifting _____
    Average nr. of procedures/day **requiring** total/partial patient lifting (**LA**) __4____

| Nr. of procedures requiring patient handling (GA + LA) | 12 | **NS** |
|---|---|---|

### DISABLED PATIENT HANDLING OPERATIONS

| Type of anesthesia | GA = Nr___3___ | | GA = Nr___5____ | |
|---|---|---|---|---|
| ☐ bed/stretcher | ☐ manual | ☐ aided | ☐ manual | ☐ aided |
| ☐ stretcher/ operating table | ☐ manual | ☐ aided | ☐ manual | ☐ aided |
| ☐ operating table / stretcher | ☐ manual | ☐ aided | ☐ manual | ☐ aided |
| ☐ stretcher / bed | ☐ manual | ☐ aided | ☐ manual | ☐ aided |
| ☐ stretcher / stretcher | ☐ manual | ☐ aided | ☐ manual | ☐ aided |
| ☐ prone / supine | ☐ manual | ☐ aided | ☐ manual | ☐ aided |
| ☐ supine / prone | ☐ manual | ☐ aided | ☐ manual | ☐ aided |
| Column score (Nr of maneuvers x Nr GA) | _6_ A | _6_ B | ____ C | _10_ D |
| Type of anesthesia | LA = Nr___4___ | | LA = Nr_____ | |
| ☐ bed / stretcher | ☐ manual | ☐ aided | ☐ manual | ☐ aided |
| ☐ stretcher / operating table | ☐ manual | ☐ aided | ☐ manual | ☐ aided |
| ☐ operating table / stretcher | ☐ manual | ☐ aided | ☐ manual | ☐ aided |
| ☐ stretcher / bed | ☐ manual | ☐ aided | ☐ manual | ☐ aided |
| ☐ stretcher / stretcher | ☐ manual | ☐ aided | ☐ manual | ☐ aided |
| ☐ prone / supine | ☐ manual | ☐ aided | ☐ manual | ☐ aided |
| ☐ supine / prone | ☐ manual | ☐ aided | ☐ manual | ☐ aided |
| Column score (Nr of maneuvers x Nr LA) | _8_ E | _8_ F | ____ G | ____ H |

DISABLED PATIENT HANDLING OPERATIONS = A+B+C+D+E+F+G+H = |__38__|

PERCENTAGE OF AIDED MANEUVERS: Sum of scores:  $\frac{B + D + F + H}{A+B+C+D+E+F+G+H}$  x 100     => |__63%__| AMPER

**FIGURE 12.8**   Completed form for operating rooms. *(continued)*

| OPERATOR TRAINING | | | | | |
|---|---|---|---|---|---|
| **TRAINING** | | | **INFORMATION** | | |
| Attended theoretical-practical course | ☐ YES | ☑ NO | Training only on how to use equipment | ☐ YES | ☐ NO |
| if YES, how many months ago? and how many hours/operator | months ____ hours ____ | | Only provided brochures on MPH | ☑ YES | ☐ NO |
| if YES, how many operators? | | | if YES, how many operators? | 18 | |
| Was EFFECTIVENESS measured and documented in writing? | | | ☐ YES | ☑ NO | |

## 2. ON-SITE INSPECTION

### EQUIPMENT FOR DISABLED PATIENT LIFTING/TRANSFERS

| EQUIPMENT DESCRIPTION | | Nr | Lack of essential requirements | Lack of adaptability to patients or environment | Lack of maintenance |
|---|---|---|---|---|---|
| **"MOBILIZER"=** power driven height-adjustable device with sliding board for patient transfer | | | YES    NO | YES    NO | YES    NO |
| **PATIENT TRANSFER UNIT"** = **wall-mounted** power driven height-adjustable device separating operating room from pre-operating holding area, featuring sliding board for patient transfer | | | YES    NO | YES    NO | YES   NO |
| Adjustable **STRETCHER** type: | | | YES   NO | YES   NO | YES   NO |
| Adjustable **STRETCHER** type: | | | YES   NO | YES   NO | YES   NO |

| EQUIPMENT DESCRIPTION | | Nr | Lack of essential requirements | Lack of adaptability to patients or environment | Lack of maintenance |
|---|---|---|---|---|---|
| SLIDING SHEETS : | | | YES   NO | YES   NO | YES   NO |
| SLIDING BOARDS : | | 2 | YES **NO** | YES **NO** | YES **NO** |
| OTHER: | | | YES   NO | YES   NO | YES   NO |

**FIGURE 12.8** *(continued)*   Completed form for operating rooms. *(continued)*

**DESCRIPTION OF STRETCHERS AND ENVIRONMENT/FURNISHINGS**

| CHARACTERISTICS AND INADEQUACY SCORE FOR STRETCHERS | SCORE | STRETCHERS | | | | | Total Nr stretchers |
|---|---|---|---|---|---|---|---|
| | | A | B | C | D | E | |
| | | Nr **4** | Nr | Nr | Nr | Nr | |
| Malfunctioning brakes | 1 | | | | | | \|_4_\| |
| Not height-adjustable | 2 | x | | | | | |
| Side flaps | 2 | | | | | | Total score |
| Needs to be partially lifted | 1 | | | | | | |
| Column score (Nr of stretchers x sum of scores) | | 8 | | | | | 8 |

Mean score **MSSTR** |_2_|, |_0_|_0_| = Total score for stretchers
                                         Total Nr stretchers

| CHARACTERISTICS AND INADEQUACY SCORE FOR ENVIRONMENT/FURNISHINGS | | Environment/furnishings | | | | | Total Nr rooms |
|---|---|---|---|---|---|---|---|
| | | Room A | Room B | Room C | Room D | Room E | |
| | | 1 | 1 | | | | |
| Operating table with side rails | 2 | | x | | | | \|_2_\| |
| Non-removable rails | 0.5 | | | | | | |
| Inadequate space for use of aids | 2 | x | | | | | Total Score |
| Column score (sum of scores) | | 2 | 2 | | | | 4 |

Mean score **MSENV** |_2_|, |_0_|_0_| = Total score
                                         Total Nr operating rooms

**FIGURE 12.8** *(continued)*   Completed form for operating rooms. *(continued)*

FONDAZIONE IRCCS CA' GRANDA
OSPEDALE MAGGIORE POLICLINICO
CLINICA DEL LAVORO - MILAN - ITALY
ERGONOMICS SECTION

**SUMMARY FOR SURGICAL UNIT**      Date _____

| Hospital EXAMPLE 3B | Surgical Unit GENERAL | Code |
|---|---|---|
| Number of procedures/day requiring patient handling (GA + LA) | | \|__12_\| NS |
| OPERATORS (staff performing manual patient handling tasks) PRESENT OVER 24 HOURS: \|_15__\|Op | | |

| FREQUENCY OF PATIENT HANDLING TASKS | A+B+C+D+E+F+G+H / \|__\|Op = 38/15 | \|__\|_2_\|,\|_5_\| F |
|---|---|---|

| EQUIPMENT FACTOR | | |
|---|---|---|
| PERCENTAGE OF AIDED PATIENT HANDLING TASKS observed \|_63%_\| AMPER | | **ERGONOMIC INADEQUACY** |
| EQUIPMENT SELDOM USED (AMPER below 50%) | | **HIGH** |
| EQUIPMENT SOMETIMES USED (AMPER TL below 90% but above (or equal to) 50%) | | **MEDIUM** ⟵ |
| EQUIPMENT ADEQUATELY USED (AMPER ≥ 90%) | | **NEGLIGIBLE** |

| EXPOSURE LEVEL | TICK EXPOSURE LEVEL | FREQUENCY OF PATIENT HANDLING TASKS (F) |
|---|---|---|
| ABSENT | NS = 0 | |
| NEGLIGIBLE | NS ≠ 0, AND AMPER ≥ 90% | \|__\|__\|,\|__\| F |
| **HIGH** | **NS ≠ 0, AND AMPER < 90%** | \|__\|_2_\|,\|_5_\| F |

**OTHER IMPORTANT ASPECTS WITH REGARD TO RISK REDUCTION**

| STRETCHERS FACTOR | | | | (MSSTR) observed: |
|---|---|---|---|---|
| Mean qualitative score observed (MSSTR) | 0.0 - 2.00 | 2.01 - 4.00 | 4.01- 6 | \|_2_\|,\|_0_\|_0_\| |
| ERGONOMIC INADEQUACY | NEGLIGIBLE | MEDIUM | HIGH | |

| ENVIRONMENT FACTOR | | | | (MSENV) observed: |
|---|---|---|---|---|
| Mean score non-ergonomic conditions observed (MSENV) | 0.0 -1.5 | 1.51 - 3 | 3.01- 4.5 | \|_2_\|,\|_0_\|_0_\| |
| ERGONOMIC INADEQUACY | NEGLIGIBLE | MEDIUM | HIGH | |

| TRAINING FACTOR | | | |
|---|---|---|---|
| Type of training | Adequate | Partly adequate | Completely inadequate |
| ERGONOMIC INADEQUACY | NEGLIGIBLE | MEDIUM | HIGH |

**FIGURE 12.8** *(continued)*    Completed form for operating rooms.

FONDAZIONE IRCCS CA' GRANDA
OSPEDALE MAGGIORE POLICLINICO
CLINICA DEL LAVORO – MILAN (ITALY)
ERGONOMICS SECTION

## PATIENT HANDLING RISK IN SURGICAL UNIT: CORRECTIVE ACTIONS INVOLVING LIFTING EQUIPMENT

### FROM SURGICAL UNIT COLLECTION SHEET

| HOSPITAL : EXAMPLE 3 C | SURGICAL UNIT : | | SURGICAL UNIT CODE: |
|---|---|---|---|
| NR. OF PROCEDURES REQUIRING PATIENT HANDLING (GA + LA): | **12** | **NS** | DATE : |

DISABLED PATIENT HANDLING OPERATIONS (manual and aided) = A+B+C+D+E+F+G+H = |__38__|

| PERCENTAGE OF AIDED PATIENT HANDLING TASKS observed (AMPER) | 24/38=63 % |
|---|---|

| EXPOSURE LEVEL | TICK EXPOSURE LEVEL | FREQUENCY OF PATIENT HANDLING TASKS (F) |
|---|---|---|
| ABSENT | NS = 0 | |
| NEGLIGIBLE | NS ≠ 0, AND AMPER ≥ 90% | \|__\|,\|__\| F |
| HIGH | NS ≠ 0, AND AMPER < 90% | \|__\|2_\|,\|_5_\| F |

| MANUAL DISABLED PATIENT HANDLING OPERATIONS | | |
|---|---|---|
| Type of anesthesia | GA = Nr__3____ | GA = Nr_____ |
| ☐ bed/stretcher | ▨ manual | ☐ manual |
| ☐ stretcher/ operating table | ☐ manual | ☐ manual |
| ☐ operating table / stretcher | ☐ manual | ☐ manual |
| ☐ stretcher / bed | ▨ manual | ☐ manual |
| ☐ stretcher / stretcher | ☐ manual | ☐ manual |
| ☐ prone / supine | ☐ manual | ☐ manual |
| ☐ supine / prone | ☐ manual | ☐ manual |
| Column score (Nr of maneuvers x Nr GA) | __6_ A | __ C |
| Type of anesthesia | LA = Nr__4____ | LA = Nr_____ |
| ☐ bed / stretcher | ▨ manual | ☐ manual |
| ☐ stretcher / operating table | ☐ manual | ☐ manual |
| ☐ operating table / stretcher | ☐ manual | ☐ manual |
| ☐ stretcher / bed | ▨ manual | ☐ manual |
| ☐ stretcher / stretcher | ☐ manual | ☐ manual |
| ☐ prone / supine | ☐ manual | ☐ manual |
| ☐ supine / prone | ☐ manual | ☐ manual |
| Column score (Nr of maneuvers x Nr LA) | __8_ E | ____ G |

NUMBER OF MANUAL DISABLED PATIENT HANDLING OPERATIONS = A+C+E+G = |__14__|

## PROPOSED SHORT-TERM CORRECTIVE ACTIONS:

— provide 2 adjustable stretchers and 2 sliding boards
— supply all staff with specific training on how to use lifting equipment

| ▨ ADJUSTABLE STRETCHERS (Nr = \|__\|_2_\| ) | | ▨ SLIDING BOARDS (Nr \|_2_\|__\| ) | |
|---|---|---|---|

### NEW PERCENTAGE OF AIDED LIFTING OPERATIONS: 38/38 = 100%

**FIGURE 12.9**  Choice of aids for risk reduction in operating rooms. *(continued)*

FONDAZIONE IRCCS CA' GRANDA
OSPEDALE MAGGIORE POLICLINICO

**PREDICTED TICK EXPOSURE LEVEL FOLLOWING ADOPTION OF LIFTING AIDS AND STAFF TRAINING**

| EQUIPMENT FACTOR | |
|---|---|
| PERCENTAGE OF AIDED PATIENT HANDLING TASKS observed \|_100%_\| AMPER | ERGONOMIC INADEQUACY |
| EQUIPMENT SELDOM USED (AMPER below 50%) | HIGH |
| EQUIPMENT SOMETIMES USED (AMPER TL below 90% but above (or equal to) 50%) | MEDIUM |
| EQUIPMENT ADEQUATELY USED (AMPER ≥ 90%) | NEGLIGIBLE ⇐ |

| EXPOSURE LEVEL | TICK EXPOSURE LEVEL | FREQUENCY OF PATIENT HANDLING TASKS (F) |
|---|---|---|
| ABSENT | NS = 0 | |
| NEGLIGIBLE | NS ≠ 0, AND AMPER ≥ 90% | \|_2_\|,\|_5_\| F ⇐ |
| HIGH | NS ≠ 0, AND AMPER < 90% | \|___\|,\|_\| F |

**OTHER IMPORTANT ASPECTS WITH REGARD TO RISK REDUCTION**

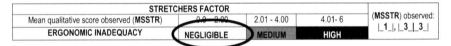

| STRETCHERS FACTOR | | | | (MSSTR) observed: |
|---|---|---|---|---|
| Mean qualitative score observed (MSSTR) | 0.0 - 2.00 | 2.01 - 4.00 | 4.01- 6 | \|_1_\|,\|_3_\|_3_\| |
| ERGONOMIC INADEQUACY | NEGLIGIBLE | MEDIUM | HIGH | |

**PROPOSED LONG TERM CORRECTIVE ACTION WITH IMPROVED QUALITY OF CARE**

Over the long term it will be necessary to verify the effective and correct use of the equipment.

**FIGURE 12.9** *(continued)*　　Choice of aids for risk reduction in operating rooms.

| No. visits/day: 150 | Patient care staff (only radiology technicians): 20 | No. operators over 24 hours: 14 |
|---|---|---|

### Brief Overview

- The radiology department consists of three rooms, each with two operators; at night there is just one room, with two operators.
- Room A receives patients directly from the Emergency Department and handles approximately 40 visits per day (all NC patients), with transfers from stretcher to x-ray table and vice versa, using a sliding board kept in room A; rooms B and C receive patients from the various wards, the day hospital, and the outpatient clinics. Of approximately 110 daily visits to both rooms, 65 are patients and 45 are disabled (10 PC and 35 NC).

### Brief Overview of Patient-Handling Activities

- All patient-handling operations are manual: NC patients are transferred from stretcher to x-ray table, and vice versa, and PC patients are transferred from wheelchair to x-ray table and vice versa.
- Equipment consists of one sliding board in room A.
- Operators have received no specific training.
- There are three stretchers, none of which is height adjustable.
- The radiology rooms do not have furnishings with side flaps or rails, and the space is adequate.
    Percentage of aided total patient lifts = 53%
    Percentage of aided partial patient lifts = 0%
    Frequency of patient-handling tasks (F): 12.1

### Remarks and Organizational Aspects

The greatest risk is due to the *absence of **adequate** equipment in the radiology department rooms, while the simplicity of the maneuvers performed there makes corrective actions easy to implement.* Patient-lifting aids must be provided in this department as soon as possible due to the high frequency (F) of patient-handling tasks.

**FIGURE 12.10**    Description of radiology department.

**DATA COLLECTION SHEET: RISK ASSESSMENT FOR MANUAL PATIENT HANDLING IN OUTPATIENT/DAY HOSPITAL SERVICES**

HOSPITAL: _____EXAMPLE 4B_____   SERVICE:___RADIOLOGY_____   Code _____   Date _____

NUMBER OF VISITS/DAY: ____150_____   OPENING HOURS: 24 HOURS  DAYS OPEN:___EVERY DAY____

## 1. INTERVIEW

| N° OF OPERATORS PERFORMING MPH: indicate the total number of operators per job category. | | | |
|---|---|---|---|
| nurses: | orderlies (ASA/OTA/OSA): | nurses aides: | other:  20 RADIOLOGY TECHNICIANS |

N° OPERATORS PERFORMING MPH TASKS OVER 3 SHIFTS: indicate number of operators on duty per shift:

| SHIFT | morning | afternoon | night |
|---|---|---|---|
| Shift hours: (from 00:00 to 00:00) | from__7 AM___to___2 PM | from__2 PM___to___9 PM | from__9 PM___to___7 A6M |
| Nr of operators over entire shift | 6 | 6 | 2 |
| (A) Total operators over entire shift = | | | 14 |

Nr OF PART-TIME OPERATORS: indicate the exact number of hours worked and calculate them as unit fractions (in relation to the overall duration of the shift).

| Nr of part-time operators present | Hours worked in shift: (from 00:00 to 00:00) | Unit fraction | (unit fraction by nr of operators) |
|---|---|---|---|
| | from_____to_____ | | |
| | from_____to_____ | | |
| (B) Total operators (as unit fractions) present by shift duration = | | | |
| TOTAL Nr OF OPERATORS ENGAGED IN MPH OVER 24 HOURS (Op): add the total number of operators present over the entire shift (A) to the total number of part-time operators (B) | | 14 | Op |

---

**AVERAGE NUMBER OF DISABLED PATIENTS (D):**

Average number/day of visits by disabled patients (D) ___85_____

(if it is difficult to quantify the average number, indicate the percentage _____%, versus the number of visits/day) of whom

Indicate the average number of totally non cooperative patients (NC) visiting the service

NC= patient who needs to be completely lifted for transfer and repositioning operations (TL)      ____75__ NC

Indicate the average number of partially cooperative patients (PC) visiting the service.

PC= patient who only needs to be partially lifted (PL)      ____10_ PC

**FIGURE 12.11**   Completed form for radiology department. *(continued)*

FONDAZIONE IRCCS CA' GRANDA
OSPEDALE MAGGIORE POLICLINICO
CLINICA DEL LAVORO – MILAN (ITALY)
ERGONOMICS SECTION

### DESCRIPTION OF MANUAL/AIDED PATIENT HANDLING OPERATIONS:

| 1)  Total lifts (TL) | NC = Nr __40_ (room A) | | NC = Nr __35_ (rooms B+C) | |
|---|---|---|---|---|
| ☐ stretcher/exam bed | ☐ manual | ▣ aided | ▣ manual | ☐ aided |
| ☐ wheelchair/exam bed | ☐ manual | ☐ aided | ☐ manual | ☐ aided |
| ☐ ward bed/exam bed | ☐ manual | ☐ aided | ☐ manual | ☐ aided |
| ☐ exam bed/ stretcher | ☐ manual | ▣ aided | ▣ manual | ☐ aided |
| ☐ exam bed/wheelchair | ☐ manual | ☐ aided | ☐ manual | ☐ aided |
| ☐ exam bed/ward bed | ☐ manual | ☐ aided | ☐ manual | ☐ aided |
| ☐ other | ☐ manual | ☐ aided | ☐ manual | ☐ aided |
| SCORE Column total<br>(Nr of maneuvers x Nr of patients) | A____ | B__80__ | C__70__ | D____ |

SUM of total lifting operations (both aided and manual) = **A+B+C+D = |__150__| SUM TL**

Percentage of aided total lifts (TL) = sum of scores:  $\dfrac{B+D}{\text{SUM TL}}$  x 100  = |__80/150= 53%__|    | ATLPER = 53% |

| 2)  Partial lifts (PL) of PC patients | PC = Nr __10__ (rooms B+C) | | PC = Nr _____ | |
|---|---|---|---|---|
| ☐ stretcher /exam bed | ☐ manual | ☐ aided | ☐ manual | ☐ aided |
| ☐ wheelchair/exam bed | ▣ manual | ☐ aided | ☐ manual | ☐ aided |
| ☐ ward bed/exam bed | ☐ manual | ☐ aided | ☐ manual | ☐ aided |
| ☐ exam bed/ stretcher | ☐ manual | ☐ aided | ☐ manual | ☐ aided |
| ☐ exam bed/wheelchair | ▣ manual | ☐ aided | ☐ manual | ☐ aided |
| ☐ exam bed/ward bed | ☐ manual | ☐ aided | ☐ manual | ☐ aided |
| ☐ other | ☐ manual | ☐ aided | ☐ manual | ☐ aided |
| Column total score<br>(Nr of maneuvers x Nr of patients) | E__20__ | F____ | G____ | H____ |

| 3)  Partial lifts (PL) of NC patients | NC = Nr _____ | | NC = Nr _____ | |
|---|---|---|---|---|
| ☐ turning over | ☐ manual | ☐ aided | ☐ manual | ☐ aided |
| ☐ pulling up in bed | ☐ manual | ☐ aided | ☐ manual | ☐ aided |
| ☐ other | ☐ manual | ☐ aided | ☐ manual | ☐ aided |
| ☐ other | ☐ manual | ☐ aided | ☐ manual | ☐ aided |
| Column total score<br>(Nr of maneuvers x Nr of patients) | X____ | Z____ | W____ | K____ |

SUM of partial lifting operations (both aided and manual) = **E+F+G+H+X+Z+W+K = |__20_| SUM PL**

Percentage of aided partial lifts (PL) = sum of scores:  $\dfrac{F+H+Z+K}{\text{SUM PL}}$  x 100  = |__0/20 = 0__|    | APLPER = 0% |

| **DISABLED PATIENT HANDLING OPERATIONS = SUM TL + SUM PL  = |__170__|** |
|---|

| OPERATOR TRAINING | | | | |
|---|---|---|---|---|
| **TRAINING** | | **INFORMATION** | | |
| Attended theoretical/practical course | ☐ YES    ▣ NO | Training only on how to use equipment | ☐ YES | ▣ NO |
| if YES, how many months ago?<br>and how many hours/operator | Months _____<br>hours _____ | Only provided brochures on MPH | ☐ YES | ☐ NO |
| if YES, how many operators? | | if YES, how many operators? | | |
| Was EFFECTIVENESS measured and documented in writing? | | ☐ YES | | ☐ NO |

**FIGURE 12.11** *(continued)*    Completed form for radiology department. *(continued)*

## 2. ON-SITE INSPECTION

### EQUIPMENT FOR DISABLED PATIENT LIFTING/TRANSFERS

| EQUIPMENT DESCRIPTION | | Nr | Lack of essential requirements | | Lack of adaptability to patients or environment | | Lack of maintenance | |
|---|---|---|---|---|---|---|---|---|
| LIFTING EQUIPMENT type: | | | YES | NO | YES | NO | YES | NO |
| Adjustable EXAM BED: | | | YES | NO | YES | NO | YES | NO |
| Adjustable STRETCHER: | | | YES | NO | YES | NO | YES | NO |
| OTHER: | | | YES | NO | YES | NO | YES | NO |

### OTHER AIDS (MINOR AIDS):

| EQUIPMENT DESCRIPTION | | Nr | Lack of essential requirements | | Lack of adaptability to patients or environment | | Lack of maintenance | |
|---|---|---|---|---|---|---|---|---|
| SLIDING SHEETS: | | | YES | NO | YES | NO | YES | NO |
| ERGONOMIC BELTS: | | | YES | NO | YES | NO | YES | NO |
| SLIDING BOARDS OR ROLLER BOARDS: | | 1 | YES | NO | YES | NO | YES | NO |
| OTHER: | | | YES | NO | YES | NO | YES | NO |

* N.B.: Attach floor plan to assess available space for more equipment and if there is an equipment storage room

### DESCRIPTION OF ROUTINELY USED STRETCHERS/WHEELCHAIRS

| CHARACTERISTICS AND INADEQUACY SCORE FOR STRETCHERS | SCORE | STRETCHERS | | | | | Total Nr stretchers |
|---|---|---|---|---|---|---|---|
| | | A | B | C | D | E | |
| | | Nr 3 | Nr | Nr | Nr | Nr | |
| Malfunctioning brakes | 1 | | | | | | \|_3_\| |
| Not height-adjustable | 2 | X | | | | | |
| Side flaps | 2 | | | | | | Total stretchers score |
| Needs to be partially lifted | 1 | | | | | | |
| Column score (Nr of stretchers x sum of scores) | 6 | | | | | | 6 |

Mean score (MSSTR) |____2____| = Total score stretchers / Total Nr stretchers

**FIGURE 12.11** *(continued)* Completed form for radiology department. *(continued)*

| WHEELCHAIR FEATURES AND INADEQUACY SCORE | Score | TYPES OF WHEELCHAIRS | | | | | | | |
|---|---|---|---|---|---|---|---|---|---|
| | | A Nr | B Nr | C Nr | D Nr | E Nr | F Nr | G Nr | |
| Malfunctioning brakes | 1 | | | | | | | | Total Nr wheelchairs |
| Non-removable armrests | 1 | | | | | | | | \|____\| |
| Cumbersome backrest | 1 | | | | | | | | Total |
| Width exceeding 70 cm | 1 | | | | | | | | Wheelchair score: |
| **Column score** (Nr of wheelchairs x sum of scores) | | | | | | | | | |

Mean score (**MSWh**) = Total wheelchairs score / total Nr wheelchairs   \|_____\| **MSWh**

## MEAN WHEELCHAIRS AND STRETCHERS INADEQUACY SCORE (MSSTR+ MSWh) =   \|____\|

### DESCRIPTION OF OUTPATIENT ENVIRONMENT/FURNISHINGS WITH EXAM ROOMS

| CHARACTERISTICS AND INADEQUACY SCORE FOR ENVIRONMENT/FURNISHINGS | Score | Environment/furnishings - Rooms | | | | | |
|---|---|---|---|---|---|---|---|
| | | Room A | Room B | Room C | Room D | Room E | Total Nr rooms |
| Free space inadequate for use of aids | 2 | | | | | | \|_3_\| |
| Exam bed not height-adjustable | 2 | | | | | | |
| Side flaps-exam bed | 1 | | | | | | |
| Part of exam bed needs to be raised manually | 1 | | | | | | |
| Patient armchair height less than 50 cm | 0,5 | | | | | | Total Score rooms |
| Door width < 85 cm | 1 | | | | | | |
| **Column score** (sum of scores) | | 0 | 0 | 0 | | | 0 |

### DESCRIPTION OF DAY HOSPITAL ENVIRONMENT/FURNISHINGS: TYPES OF ROOMS

| CHARACTERISTICS AND INADEQUACY SCORE FOR WARD ROOMS | Score | Nr rooms with Nr___ beds | Nr rooms with Nr___ beds | Nr rooms with Nr___ beds | Nr rooms with Nr___ beds | Nr rooms with Nr___ beds | |
|---|---|---|---|---|---|---|---|
| Space between beds or between bed and wall less than 90 cm | 2 | | | | | | Total Nr DH rooms |
| Space between foot of bed and wall less than 120 cm | 2 | | | | | | \|____\| |
| Unsuitable bed that needs to be partially lifted | 1 | | | | | | |
| Space between bed and floor less than 15 cm | 2 | cm | cm | cm | cm | cm | Total score DH rooms: |
| Patient armchair height less than 50 cm | 0,5 | | | | | | |
| **Column score** (Nr of rooms x sum of scores) | | | | | | | |

Mean environment inadequacy score (**MSENV**) = $\dfrac{\text{total room score + total DH room score}}{\text{total Nr of rooms + total Nr of DH rooms}}$

### NOTES

_____

_____

_____

**FIGURE 12.11** *(continued)*    Completed form for radiology department. *(continued)*

FONDAZIONE IRCCS CA' GRANDA
Ospedale Maggiore Policlinico
CLINICA DEL LAVORO  (MILAN ITALY)
EPM (EPM UNIT - CEMOC)

### SUMMARY OF OUTPATIENT/DAY HOSPITAL SERVICES

| Hospital EXAMPLE 4B | Service RADIOLOGY | | Code |
|---|---|---|---|
| Nr visits/day __150___   Average number/day of visits by disabled patients (D) _85_ | | OPERATORS (Op) | \|_14_\| |
| FREQUENCY OF PATIENT HANDLING TASKS | SUM TL + SUM PL /Op = 170/14 | | \|_1_\|_2_\|,\|_1_\| F |
| Nr totally non-cooperative patients NC __75_____     Nr partially cooperative patients PC __10_____ | | | |

| "TOTAL LIFTING DEVICES" FACTOR | ERGONOMIC INADEQUACY |
|---|---|
| PERCENTAGE OF TOTAL AIDED LIFTS \|__\| ATLPER | |
| EQUIPMENT SELDOM USED (ATLPER below 50%) | HIGH |
| EQUIPMENT SOMETIMES USED (ATLPER below 90% but above (or equal to) 50%) | MEDIUM |
| EQUIPMENT ADEQUATELY USED (ATLPER ≥ 90%) | NEGLIGIBLE |

| "PARTIAL LIFTING DEVICES" FACTOR | ERGONOMIC INADEQUACY |
|---|---|
| PERCENTAGE OF AIDED PARTIAL LIFTS \|__\| APLPER | |
| EQUIPMENT SELDOM USED (APLPER below 90%) | HIGH |
| EQUIPMENT ADEQUATELY USED (APLPER ≥ 90%) | NEGLIGIBLE |

| EXPOSURE LEVEL | TICK EXPOSURE LEVEL | FREQUENCY OF PATIENT HANDLING TASKS (F) |
|---|---|---|
| ABSENT | D = 0 | |
| NEGLIGIBLE | ATLPER ≥ 90% AND APLPER ≥ 90% | \|__\|,\|__\| F |
| MEDIUM | ATLPER ≥ 90% AND APLPER < 90% | \|__\|,\|__\| F |
| HIGH | ATLPER < 90% | \|_1_\|_2_\|,\|_1_\| F |

### OTHER IMPORTANT ASPECTS WITH REGARD TO RISK REDUCTION

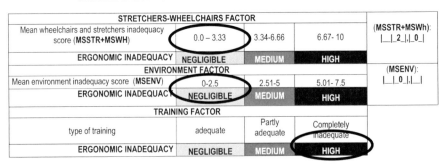

| STRETCHERS-WHEELCHAIRS FACTOR | | | | (MSSTR+MSWh): |
|---|---|---|---|---|
| Mean wheelchairs and stretchers inadequacy score (MSSTR+MSWH) | 0.0 – 3.33 | 3.34-6.66 | 6.67- 10 | \|__\|_2_\|,\|_0_\| |
| ERGONOMIC INADEQUACY | NEGLIGIBLE | MEDIUM | HIGH | |
| ENVIRONMENT FACTOR | | | | (MSENV): |
| Mean environment inadequacy score (MSENV) | 0-2.5 | 2.51-5 | 5.01- 7.5 | \|__\|_0_\|,\|__\| |
| ERGONOMIC INADEQUACY | NEGLIGIBLE | MEDIUM | HIGH | |
| TRAINING FACTOR | | | | |
| type of training | adequate | Partly adequate | Completely inadequate | |
| ERGONOMIC INADEQUACY | NEGLIGIBLE | MEDIUM | HIGH | |

**FIGURE 12.11** *(continued)*   Completed form for radiology department.

**PATIENT HANDLING RISK IN** OUTPATIENT/DAY HOSPITAL SERVICES: **CORRECTIVE ACTIONS INVOLVING LIFTING EQUIPMENT**

**FROM OUTPATIENT/DAY HOSPITAL SERVICES COLLECTION SHEET**

| HOSPITAL : **EXAMPLE 4C** | SERVICE : **RADIOLOGY** | DATE: |
|---|---|---|
| NUMBER OF VISITS/DAY: 150 | OPENING HOURS: 24 HOURS | DAYS OPEN: Every days |
| % OF AIDED TOTAL LIFTING OPERATIONS (% ATL) | | 53 % |
| % OF AIDED PARTIAL LIFTING OPERATIONS (% APL) | | 0 % |

| EXPOSURE LEVEL | TICK EXPOSURE LEVEL | FREQUENCY OF PATIENT HANDLING TASKS (F) |
|---|---|---|
| ABSENT | D = 0 | |
| NEGLIGIBLE | ATLPER ≥ 90%AND APLPER ≥ 90% | \|_\|_\|,\|_\| F |
| MEDIUM | ATLPER ≥ 90% AND APLPER < 90% | \|_\|_\|,\|_\| F |
| HIGH | ATLPER < 90% | \|_1_\|_2_\|,\|_1_\| F |

| MANUAL DISABLED PATIENT HANDLING OPERATIONS | | |
|---|---|---|
| **Total lifts (TL)** | NC = Nr __35 (Rooms B+C)____ | |
| ☐ stretcher/exam bed | ☐ manual | ☐ manual |
| ☐ wheelchair/exam bed | ☐ manual | ☐ manual |
| ☐ ward bed/exam bed | ☐ manual | ☐ manual |
| ☐ exam bed/ stretcher | ☐ manual | ☐ manual |
| ☐ exam bed/wheelchair | ☐ manual | ☐ manual |
| ☐ exam bed/ward bed | ☐ manual | ☐ manual |
| ☐ other | ☐ manual | ☐ manual |
| SCORE Column total (Nr of maneuvers x Nr of patients) | A___ | C_70 |
| **Partial lifts (PL) of PC patients** | PC = Nr _10 _(Rooms B+C)__ | |
| ☐ stretcher/exam bed | ☐ manual | ☐ manual |
| ☐ wheelchair/exam bed | ☐ manual | ☐ manual |
| ☐ ward bed/exam bed | ☐ manual | ☐ manual |
| ☐ exam bed/ stretcher | ☐ manual | ☐ manual |
| ☐ exam bed/wheelchair | ☐ manual | ☐ manual |
| ☐ exam bed/ward bed | ☐ manual | ☐ manual |
| ☐ other | ☐ manual | ☐ manual |
| SCORE Column total (Nr of maneuvers x Nr of patients) | E_20_ | G___ |
| **Partial lifts (PL) of NC patients** | NC = Nr _____ | |
| ☐ turning over | ☐ manual | ☐ manual |
| ☐ pulling up in bed | ☐ manual | ☐ manual |
| ☐ other | ☐ manual | ☐ manual |
| ☐ other | ☐ manual | ☐ manual |
| SCORE Column total (Nr of maneuvers x Nr of patients) | X___ | W___ |
| NUMBER OF MANUAL DISABLED PATIENT HANDLING OPERATIONS = A+C+E+G+X+W = \|_90\| | | |

**FIGURE 12.12** Choice of aids for risk reduction in radiology department. *(continued)*

**PROPOSED SHORT-TERM CORRECTIVE ACTIONS**

- Acquisition of 2 additional sliding boards (for room B and C). This will allow NC patient handling operations to be aided.
- Supply all staff whit specific training on how to how to use lifting equipement.

| ☐ ADJUSTABLE STRETCHERS (Nr = \|_\|_\| ) | | ☐ SLIDING BOARDS **(Nr \|2\| )** | |

## PREDICTED TICK EXPOSURE LEVEL FOLLOWING ADOPTION OF LIFTING AIDS AND STAFF TRAINING

| % OF AIDED TOTAL LIFTING OPERATIONS (% ATL) = 150/150 | 100 % |
|---|---|
| % OF AIDED PARTIAL LIFTING OPERATIONS (% APL) = 0/20 | 0 % |

| EXPOSURE LEVEL | TICK EXPOSURE LEVEL | FREQUENCY OF PATIENT HANDLING TASKS (F) |
|---|---|---|
| ABSENT | D = 0 | |
| NEGLIGIBLE | ATLPER ≥ 90%AND APLPER ≥ 90% | \| \| \|\| \|F |
| MEDIUM | ATLPER ≥ 90% AND APLPER < 90% | \|1\|2\|,\|1\| F |
| HIGH | ATLPER < 90% | \|_\|_\|_\| F |

**PROPOSED LONG-TERM CORRECTIVE ACTIONS**

Aids are still required for transferring patients accessing the service by wheelchair. Options include:

- A height-adjustable wheelchair with small sliding sheet or board
  Or alternatively (and preferably)
- The re-organization of incoming patients so that all patients arrive (and are transported) on stretchers.
- There is an on-going need for adequate specific risk training.

## PREDICTED TICK EXPOSURE LEVEL FOLLOWING LONG TERM CORRECTIVE ACTION

| EXPOSURE LEVEL | TICK EXPOSURE LEVEL | FREQUENCY OF PATIENT HANDLING TASKS (F) |
|---|---|---|
| ABSENT | D = 0 | |
| NEGLIGIBLE | ATLPER ≥ 90%AND APLPER ≥ 90% | \|_1_\|_2_\|,\|_1_\| F |
| MEDIUM | ATLPER ≥ 90% AND APLPER < 90% | \|_\|,\|_\| F |
| HIGH | ATLPER < 90% | \|_\|_\|_\| F |

**FIGURE 12.12** *(continued)*  Choice of aids for risk reduction in radiology department.

# References

Andersson, G. B.J. 1999. Epidemiological features of chronic low-back pain. *Lancet* 354 (9178): 581–585.

Ariens, G. A. M., van Mechelen, W., Bongers, P. M., Buoter, L. M., van der Wal, G. 2000. Physical risk factors for neck pain. *Scandinavian Journal of Work and Environmental Health* 26(1): 7–19.

Atlas, S. J., Wasiak, R., van den Ancker, M., Webster, B., Pransky, G. 2004. Primary care involvement and outcomes of care in patients with a workers' compensation claim for back pain. *Spine* 29(9): 1041–1048.

Baldasseroni, A., Abrami, V., Arcangeli, G., Capelli, V., Fiorito, M., Guarducci, L., Sommani, L., Tartaglia, R. 2005. Studio longitudinale per la valutazione dell'efficacia di misure preventive in una popolazione di operatori sanitari esposta al rischio di movimentazione manuale di pazienti. *Giornale Italiano de Medicina del Lavoro ed Ergonomia* 27(1): 101–105.

Baldasseroni, A., Tartaglia, R., Sgarrella, C., Carnevale, F. 1998. Frequency of lumbago in a cohort of nursing students. *La Medicina del Lavoro* 89(3): 242–253.

Battevi, N., Consonni, D., Ricci, M.G., Menoni, O., Occhipinti, E., Colombini, D. 1999. L'applicazione dell'indice sintetico di esposizione nella movimentazione manuale pazienti: prime esperienze di validazione. *La Medicina del Lavoro* 90(2): 256–275.

Battevi, N., Menoni, O., Alvarez-Casado, E. 2012. Screening of patient manual handling risk using the MAPO method. *La Medicina del Lavoro* 103(1): 37–48. (Italian)

Battevi, N., Menoni, O., Ricci, M.G., Cairoli, S. 2006. MAPO index for risk assessment of patient manual handling in hospital wards: A validation study. *Ergonomics* 49(7): 671–687.

Battié, M. C. 1990. Anthropometric and clinical measures as predictors of back pain complaints in industry: A prospective study. *Journal of Spinal Disorders* 3(3): 195–204.

Beruffi, M., Mossini, M., Zamboni, R. 1999. The assessment of exposure to the risk of the manual lifting of patients and the results of a clinical study in the rest homes of the Mantua area. *La Medicina del Lavoro* 90(2): 291–307.

Bongers, P. M., de Winter, C. R., Kompier, M. A., Hildebrandt, V. H. 1993. Psychosocial factors at work and musculoskeletal disease. *Scandinavian Journal of Work and Environmental Health* 19(5): 297–312.

Borg, G. 1998. *Perceived Exertion and Pain Scales*. Champaign, IL: Human Kinetics.com.

Bos, E., Krol, B., van der Star, L., Groothoff, J. 2007. Risk factors and musculoskeletal complaints in non-specialized nurses, IC nurses, operation room nurses, and x-ray technologists. *International Archives of Occupational and Environmental Health* 80(3): 198–206. Epub 2006 Jun 24.

Buerhaus, P. I., Staiger, D. O., Auerbach, D. I. 2000. Implications of an aging registered nurse workforce. *Journal of American Medical Association* 283: 2948–2954.

Burdof, A., Koppelaar, E., Bradley, E. 2013. Assessment of the impact of lifting devices use on low back pain and musculoskeletal injury claims among nurses. *Occupational and Environmental Medicine* 70: 491–497.

Cady, L. D., Jr., Thomas, P. C., Karwasky, R. J. 1985. Program for increasing health and physical fitness of fire fighters. *Journal of Occupational Medicine* 27(2): 110–114.

Camerino, D., Cesana, G. C., Molteni, G., De Vito, G., Evaristi, C., Latocca, R. 2001. Job strain and musculoskeletal disorders of Italian nurses. *Occupational Ergonomics* 2: 215–223.

Camerino, D., Lusignani, M., Conway, P. M., Bertazzi, P. A., Gruppo N. 2004. Intention to leave the nursing profession. *La Medicina del Lavoro* 95(5): 354–364.

Camerino, D., Molteni, G., Finotti, S., Capietti, M., Molinari, M., Cotroneo, L., Morselli, G. 1999. La prevenzione del rischio da movimentazione manuale dei pazienti: la componente psicosociale. *La Medicina del Lavoro* 90: 412–427.

Caruso, C. C., Waters, T. R. 2008. A review of work schedule issues and musculoskeletal disorders with an emphasis on the healthcare sector. *Industrial Health* 46: 523–534.

Chaffin, D. B., Andersson, G. B. J., Martin, B. J. 2006. *Occupational Biomechanics, 4th edition*. John Wiley & Sons, New York.

Collins, J. W., Menzel, N. N. 2006. *Safe Patient Handling and Movement: Practical Guide for Health Care Professionals*. New York: Springer Publishing, 1–26.

Collins, J. W., Wolf, L., Bell, J., Evanoff, B. 2004. An evaluation of a "best practices" musculoskeletal injury prevention program in nursing homes. *Injury Prevention* 10: 206–211.

Colombini, D., Menoni, O., Occhipinti, E., Battevi, N., Ricci, M. G., Cairoli, S., Sferra, G., Cimaglia, G., Missere, M., Draicchio, F., Papale, A., Di Loreto, G., Ubiali, E., Bertolini, C., Piazzini, D. B. 2005. Criteri per la classificazione di casi di malattia da sovraccarico biomeccanico degli arti superiori nell'ambito della medicina del lavoro. Documento di consenso di un gruppo di lavoro nazionale. *La Medicina del Lavoro* 96, suppl 2.

Colombini, D., Occhipinti, E., Alvarez, C., Waters, T. 2013. *Manual Lifting: A Guide to the Study of Simple and Complex Lifting Tasks*. Taylor & Francis Group, Boca Raton, FL: CRC Press.

Colombini, D., Riva, D., Lue, F., Nava, C., Petri, A., Basilico, S., Linzalata, M., Morselli, G., Cotroneo, L., Ricci, M. G., Menoni, O., Battevi, N. 1999. Primi dati epidemiologici di esperienze nazionali sugli effetti clinici negli operatori sanitari addetti alla movimentazione manuale di pazienti nei reparti di degenza. *Med. Lav.* 90(2): 201–228.

Corona, G., Amedei, F., Miselli, F., Padalino, M. P., Tibaldi, S., Franco, G. 2005. Associazione fra fattori relazionali e organizzativi e insorgenza di patologia muscoloscheletrica negli operatori sanitari. *Giornale Italiano de Medicina del Lavoro ed Ergonomia* 27: 2.

Corona, G., Monduzzi, G., Minerva, M., Amedei, F., Franco, G. 2004. Individual, ergonomic and psychosocial risk factors affect musculoscheletal disorders in nurses, physiotherapist and VDU users. *Giornale Italiano de Medicina del Lavoro ed Ergonomia* 26 (suppl): 201–202.

Croft, P. R., Lewis, M., Papageorgiou, A. C., Thomas, E., Jayson, Mi., Macfarlane, G. J., Silman, A. J. 2001. Risk factors for neck pain: A longitudinal study in the general population. *Pain* 93(3): 317–325.

Daynard, D., Yassi, A., Cooper, J. E., Tate, R., Norman, R., Wells, R. 2001. Biomechanical analysis of peak and cumulative spinal loads during simulated patient-handling activities: A substudy of randomized controlled trial to prevent lift and transfer injury of health care workers. *Applied Ergonomics* 32: 199–214.

Dehlin, O., Jäderberg, E. 1982. Perceived exertion during patient lifts. An evaluation of the importance of various factors for the subjective strain during lifting and carrying of patients. A study at a geriatric hospital. *Scandinavian Journal of Rehabilitative Medicine* 14(1): 11–20.

Devereux, J. J., Vlachonikolis, I. G., Bucle, P. W. 2001. Epidemiological study to investigate potential interaction between physical and psychosocial factors at work that may increase the risk of symptoms of musculoskeletal disorder of the neck and upper limb. Reperibile su: www.occenvmed.com

Edlich, R. F., Hudson, M. A., Buschbacher, R. M., Winters, K. L., Britt, L. D., Cox, M. J., Becker, D. G., McLaughlin, J. K., Gubler, K. D., Zomerschoe, T. S., Latimer, M. F., Zura, R. D., Paulsen, N. S., Long, W. B. 3rd, Brodie, B. M., Berenson, S., Langenburg, S. E., Borel, L., Jenson, D. B., Chang, D. E., Chitwood, W. R. Jr., Roberts, T. H., Martin, M. J., Miller, A., Werner, C. L., Taylor, P. T. Jr., Lancaster, J., Kurian, M. S., Falwell,

J. L. Jr., Falwell, R. J. 2005. Devastating injuries in healthcare workers: Description of the crisis and legislative solution to the epidemic of back injury from patient lifting. *Journal of Long Term Effects of Medical Implants* 15(2): 225–241.

Elfering, A., Grebner, S., Gerber, H., Semmer, N. K. 2008. Workplace observation of work stressors, catecholamines and musculoscheletal pain among male employees. *Scandinavian Journal of Environmental Health* 34(5): 337–344.

EN 1005-1:2001. Safety of machinery—Human physical performance. Part 1: Terms and definitions.

EN 1005-2:2003, Safety of machinery—Human physical performance. Part 2: Manual handling of machinery and component parts of machinery.

EN 1005-3:2002. Safety of machinery—Human physical performance. Part 3: Recommended force limits for machinery operation.

EN 1005-4:2005. Safety of machinery—Human physical performance. Part 4: Evaluation of working postures and movements in relation to machinery.

EN 1005-5:2007. Safety of machinery—Human physical performance. Part 5: Risk assessment for repetitive handling at high frequency.

EN 12182:1999. General requirements for assistive products for person with disability.

EN 1441:1998. Medical devices. Risk analysis.

EN 1970:2000. Adjustable beds for disabled persons—Requirements and test methods.

EN 614-1:2006+A1:2009. Safety of machinery—Ergonomic design principles. Part 1: Terminology and general principles.

EN 614-2:2000+A1:2008. Safety of machinery—Ergonomic design principles. Part 2: Interactions between the design of machinery and work tasks.

Engkvist, I. L., Hjelm, E. W., Hagberg, M., Menckel, E., Ekenvall, L. 2000. Risk indicators for reported over-exertion back injuries among female nursing personnel. *Epidemiology* 11(5): 519–522.

Eriksen, H. R., Ihlebaek, C., Jansen, J. P., Burdorf, A. 2006. The relations between psychosocial factors at work and health status among workers in home care organizations. *International Journal of Behavioral Medicine* 13(3): 183–192.

European Agency for Safety and Health at Work. 2007a. Prevention of work-related MSDs in practice.

————. 2007b. Work-related musculoskeletal disorders: Back to work report.

European Foundation for the Improvement of Living and Working Conditions. 2005. 2005—Annual Report. Retrieved from http://www.eurofound.europa.eu/publications/htmlfiles/ef0513.htm.

EWCS (European Working Conditions Survey). 2007. http://www.eurofound.europa.eu/ewco/

Feldstein, A., Vollmer, W., Valanis, B. 1990. Evaluating the patient-handling tasks of nurses. *Journal of Occupational Medicine* 32(10): 1009–1013.

Ferguson, S. A., Marras, W. S. 1997. A literature review of low back disorder surveillance measures and risk factors. *Clinical Biomechanics* (Bristol, Avon) 12(4): 221–226.

Feyer, A. M., Herbison, P., Williamson, A. M., de Silva, I., Mandryk, J., Hendrie, L., Hely, M. C. 2000. The role of physical and psychological factors in occupational low back pain: A prospective cohort study. *Occupational and Environmental Medicine* 57(2): 116–120.

Folletti, I. et al. 2005. Prevalence and determinants of low back pain in hospital worker. *Giornale Italiano de Medicina del Lavoro ed Ergonomia* 27(3): 359–361. (Italian)

Fray, M. 2010. A comprehensive evaluation of outcomes from patient handling interventions. PhD thesis. Loughborough Design School, Loughborough University.

Frigo, C. 1999. Instrumental studies in laboratories dedicated to the examination of disk overload in health workers. *La Medicina del Lavoro* 90(2): 117–130. Fujishiro, K., Weaver, J. L., Heaney, C. A., Hamrick, C. A., Marras, W. S. 2005. The effect of ergonomic interventions in healthcare facilities on musculoskeletal disorders. *American Journal of Industrial Medicine* 48(5): 338–347.

Garg, A. 2006. Prevention of injuries in nursing homes and hospitals. *Proceedings of the IEA 2006 Congress,* ed. R. N. Pikaar, E. A. P. Koningsveld, P. J. M. Settels. Elsevier Ltd.

Giannandrea, F., Marini Bettolo, P., Iezzi, D. F., Chirico, F., Bernardini, P. 2004. Rischio di lombalgia in operatori sanitari valutato con l'Oswestry Disability Index (ODI). *Giornale Italiano de Medicina del Lavoro ed Ergonomia* 26: 4 Suppl.

Global burden of diseases, injuries, and risk factors study. 2013. http://www.healthmetricsandevaluation.org/gbd/2013

Granata, K. P., England, S. A. 2006. Stability of dynamic trunk movement. *Spine* 31(10): 271–276.

Granata, K. P., Marras, W. S. 1995. The influence of trunk muscle coactivity on dynamic spinal loads. *Spine* 20(8): 913–919.

Guardini, I. Deroma, L. Salmaso, D. Palese, A. 2011. Assessing the trend in the aging of the nursing staff at two hospitals of the Friuli Venezia Giulia region: Application of a deterministic mathematical model. *Giornale Italiano de Medicina del Lavoro ed Ergonomia* 33(1): 55–62.

Harkness, E. F., Macfarlane, G. J., Nahit, E. S., Silman, A. J., McBeth, J. 2003. Mechanical and psychosocial factors predict new onset shoulder pain: A prospective cohort study of newly employed workers. *Occupational and Environmental Medicine* 60(11): 850–857.

Hashemi, L., Webster, B. S., Clancy, E. A. 1998. Trends in disability duration and cost of workers' compensation low back pain claims. *Journal of Occupational and Environmental Management* 40: 12.

Hignett, S. 1996. Work-related back pain in nurses. *Journal of Advanced Nursing* 23(6): 1238–1246.

———. 2001. Embedding ergonomics in hospital culture: Top-down and bottom-up strategies. *Applied Ergonomics* 32: 61–69.

———. 2003. Intervention strategies to reduce musculoskeletal injuries associated with handling patients: A systematic review. *Occupational and Environmental Medicine* 60(9): E6.

Hignett, S., Fray, M., Battevi, N., Occhipinti, E., Menoni, O., Tamminen-Peter, L., Waaijer, E., Knibbe, H., Jager, M. 2014. International consensus on manual handling of people in the healthcare sector: Technical report ISO/TR 12296. International *Journal of Industrial Ergonomics* 44: 191–195.

Hignett, S., Keen, E. 2005. How much space is needed to operate a mobile and an overhead patient hoist? *Professional Nurse* 20(7): 40–42.

Hignett, S., McAtamney, L. 2000. Rapid entire body assessment. *Applied Ergonomics* 31: 201–205.

Hooftman, W. E. 2004. Gender differences in the relations between work-related physical and psychosocial risk factors and musculoskeletal complaints. *Scandinavian Journal of Work and Environmental Health* 30(4): 261–278.

Hooftman, W. E., van der Beek, A. J., Bongers, P. M., van Mechele, W. 2009. Is there a gender difference in the effect of work-related physical and psychosocial risk factors on musculoskeletal symptoms and related sickness absence? *Scandinavian Journal of Work and Environmental Health* 35(2): 85–95.

Hoogendoorn, W. E., Bongers, P. M., de Vet, H. C., Ariens, G. A., van Mechelen, W., Bouter, L. M. 2002 High physical work load and low job satisfaction increase the risk of sickness absence due to low back pain: results of a prospective cohort study. *Occupational and Environmental Medicine* 59(5): 323–328.

Hoogendoorn, W. E., van Poppel, M. N., Bongers, P. M., Koes, B. W., Bouter, L. M. 1999. Physical load during work and leisure time as risk factors for back pain. *Scandinavian Journal of Work and Environmental Health* 25(5): 387–403. Retrieved from www.bls.gov/iif/oshwc/osh/case/ostb1039.pdf

Hoozemans, M. J., van der Beek, A. J., Fring-Dresen, M. H., van der Woude, L. H., van Dijk, F. J. 2002. Low-back and shoulder complaints among workers with pushing and pulling tasks. *Scandinavian Journal of Work and Environmental Health* 28(5): 293–303.

Howard, N., Adams, D. 2010. An analysis of injuries among home health care workers using the Washington state workers' compensation claims database. *Home Health Care Services Quarterly* 29(2): 55–74.

Hyun, K., Dropkin, J., Spaeth, K., Smith, F., Moline, J. 2011. Patient handling and musculoskeletal disorders among hospital workers: Analysis of 7 years of institutional workers' compensation claims data. *American Journal of Industrial Medicine* 55: 683–690.

Iles, R. A., Davidson, M., Taylor, N. F. 2008. Psychosocial predictors of failure to return to work in non-chronic non-specific low back pain: A systematic review. *Occupational and Environmental Medicine* 65: 507–517.

ISO 10535. 2006, Hoists for the transfer of disabled persons. Requirements and test methods.

ISO 11295. 2010. Classification and information on design of plastics piping systems used for renovation.

ISO 12100. 2010. Safety of machinery—Basic concepts, general principles for design.

ISO 14121-1. 2007. Safety of machinery—Risk assessment. Part 1: Principles.

ISO/TR 12296. 2012. Ergonomics—Manual handling of people in the health care sector.

Jäger, M., Jordan, C., Theilmeier, A., Luttmann, A., Dolly Group. 2007. Spinal-load analysis of patient-transfer activities. In *Advances in Medical Engineering,* ed. T. M. Buzug, D. Holz, S. Weber, J. Bongartz, M. Kohl-Bareis, p. 273–278. Berlin: Springer.

———. 2010. Lumbar-load quantification and overload-risk prevention for manual patient handling—The Dortmund approach. In *Proceedings of 8th International Conference on Occupational Risk Prevention,* Ed. P. Mondelo, W. Karwowski, K. Saarela, A. Hale, E. Occhipinti. ORP2010, CD-Rom (9 pp.)

Jang, R. et al. 2007. Biomechanical evaluation of nursing tasks in a hospital setting. *Ergonomics* 50(11): 1835–1855.

Johnsson, C., Kjellberg, K., Kjellberg, A., Lagerström, M. 2004. A direct observation instrument for assessment of nurses' patient transfer technique (DINO). *Applied Ergonomics* 35(6): 591–601.

Karhu, O., Kansi, P., Kuorinka, I. 1977. Correcting working postures in industry: A practical method for analysis. *Applied Ergonomics* 8(4): 199–201.

Kjellberg, K., Johnsson, C., Proper, K., Olsson, E., Hagberg, M. 2000. An observation instrument for assessment of work technique in patient transfer tasks. *Applied Ergonomics* 31(2): 139–150.

Knibbe, N. E., Knibbe, J. J. 1996. Postural load of nurses during bathing and showering of patients: Results of a laboratory study. LOCOmotion, Professional Safety USA.

Knibbe, J. J., Friele, R. D. 1999. The use of logs to assess exposure to manual handling of patients, illustrated in an intervention study in home care nursing. *International Journal of Industrial Ergonomics* 24: 445–454.

Knibbe, J. J., Knibbe, N. E. 2006. Monitoring the effects of the ergonomics covenants for workers in Dutch health care. In *Proceedings of the XV Triennial Congress of the International Ergonomics Association, Meeting Diversity in Ergonomics,* ed. R. N. Pikaar, E. A. P. Konigsveld, P. J. M. Settels, 11–14, Maastricht, Netherlands.

Knibbe, J. J., Van Panhuys, W., Van Vught, W., Waaijer, E. M., Hooghiemstra, F. 2008. *Handbook of Transfers*, Diligent: UK.

Knowles, M., Holton, E. F. III, Swanson, R. A. 2008. *Quando l'Adulto Impara. Andragogia e Sviluppo della Persona,* Milano: Franco Angeli.

Kompier, M., van der Beek, A. J. 2008. Psychosocial factors at work and musculoskeletal disorders. *Scandinavian Journal of Work and Environmental Health* 34(5): 323–325.

Lagerström, M., Hansson, T., Hagberg, M. 1998. Work-related low-back problems in nursing. *Scandinavian Journal of Work and Environmental Health* 24(6): 449–464.

Larese, F., Fiorito, A. 1994. Musculoskeletal disorders in hospital nurses: A comparison between two hospitals. *Ergonomics* 37(7): 1205–1211.

Linton, S. J. 2000. A review of psychological risk factors in back and neck pain. *Spine* 25(9): 1148–1156.

———. 2001. Occupational psychological factors increase the risk for back pain: A systematic review. *Journal of Occupational Rehabilitation* 11(1): 53–66.

Lipscomb, J. A., Trinkoff, A. M., Geiger-Brown, J., Brady, B. 2002. Work-schedule characteristics and reported musculoskeletal disorders of registered nurses. *Scandinavian Journal of Work and Environmental Health* 28(6): 394–401.

Lorusso, A., Bruno, S., L'Abbate, N. 2007. A review of low back pain and musculoskeletal disorders among Italian nursing personnel. *Industrial Health* 45(5): 637–644.

Luime, J. J., Kuiper, J. I., Koes, B. W., Verhaar, J. A., Miedema, H. S., Burdorf, A. 2004. Work-related risk factors for the incidence and recurrence of shoulder and neck complaints among nursing-home and elderly-care workers. *Scandinavian Journal of Work and Environmental Health* 30(4): 279–286.

Magora, A. 1970. Investigation of the relation between low back pain and occupation. *Industrial Medicine and Surgery* 39: 504–510.

Marena, C., Gervino, D., Pistorio, A., Azzaretti, S., Chiesa, P., Lodola, L., Marraccini, P. 1997. Epidemiologic study on the prevalence of low back pain in health personnel exposed to manual handling tasks. *Giornale Italiano de Medicina del Lavoro ed Ergonomia.* 19(3): 89–95.

Marras, W. S. 2008. *The Working Back. A Systems View.* Chichester: Wiley-Interscience.

Marras, W., Davies, K., Kirking, B., Bertsche, P. 1999. A comprehensive analysis of low-back disorder risk and spinal loading during the transferring and repositioning of patients using different techniques. *Ergonomics* 42(7): 904–926.

Marras, W. S., Kermit, G. D., Heaney, C. A., Maronitis, A. B., Allread, W. G. 2000. The influence of psychosocial stress, gender and personality on mechanical loading of the lumbar spine. *Spine* 25(23): 3045–3054.

Marras, W. S., Knapik, G. G., Ferguson, S. 2009. Lumbar spine forces maneuvering of ceiling-based and floor-based patient transfer devices. *Ergonomics* 52–53: 384–397.

Marras, W. S., Parakkat, J., Chany, A. M., Yang, G., Burr, D., Lavender, S. A. 2006. Spine loading as a function of lift frequency, exposure duration, and work experience. *Clinical Biomechanics (Bristol, Avon)* 21(4): 345–352.

Martinelli, S., Artioli, G., Vinceti, M., Bergomi, M., Bussolanti, N., Camellini, R., Celotti, P., Capelli, P., Roccato, L., Gobba, F. 2004. Low back pain risk in nurses and its prevention. *Prof Inferm* 57(4): 238–242. (Italian)

Maso, S., Furno, M., Vangelista, T., Cavedon, F., Musilli, L., Saia, B. 2003. Musculoskeletal diseases among a group of geriatric residence workers. *Giornale Italiano de Medicina del Lavoro ed Ergonomia* 25 Suppl. (3): 194–195.

Mastrangelo, G. 2008. La percezione dei rischi nei lavoratori del Veneto. Atti del Convegno Sicurezza e salute sul lavoro nelle strutture sanitarie del Veneto: Prospettive di integrazione fra i sistemi di gestione. Venezia. Retrieved from www.intranet.safetynet.it/webeditor/3/1/intranet/conv_work/atti_22_05_08.htm

Mehlum, I. S., Kristensen, P., Kjuus, H., Wergeland, E. 2008. Are occupational factors important determinants of socioeconomic inequalities in musculoskeletal pain? *Scandinavian Journal of Work and Environmental Health* 34(4): 250–259.

Menoni, O., Ricci, M. G., Panciera, D., Battevi, N., Colombini, D., Occhipinti, E., Greco, A. 1999. La movimentazione manuale dei pazienti nei reparti di degenza delle strutture sanitarie: Valutazione del rischio, sorveglianza sanitaria e strategie preventive. *La Medicina del Lavoro* 90, 2, numero monografico.

Miranda, H. 2002. Individual factors, occupational loading and physical exercise as predictors of sciatic pain. *Spine* 27(10): 1102–1109.

Miranda, H., Punnet, L., Viikari-Juntura, E., Heliovaara, M., Knekt, P. 2008. Physical work and chronic shoulder disorder. Results of prospective population-based study. *Annals of Rheumatic Diseases* 67(2): 218–223.

Myers, D., Silverstein, B., Nelson, N. A. 2002. Predictors of shoulder and back injuries in nursing home workers: A prospective study. *American Journal of Industrial Medicine* (41) 6: 466–476.

NCHS (National Center for Health Statistics). 2006. Health. *United States, 2006: With Chartbook on Trends in the Health of Americans with Special Feature on Pain.* Atlanta: CDC.

Nelson, A. 2009. *Safe Patient Handling and Movement.* New York: Springer.

Nelson, A., Matz, M., Chen, F., Siddharthan, K., Lloyd, J., Fragala, G. 2006. Development and evaluation of a multifaceted ergonomics program to prevent injuries associated with patient handling tasks. *International Journal of Nursing Studies* 43(6): 717–733.

NIOSH. 1981. Work Practices Guide for Manual Lifting; NIOSH technical report. Publication no. 81–122. US Dept. Health & Human Services.

NIOSH, CDC. 1997. *Musculoskeletal Disorders and Workplace Factors. A Critical Review of Epidemiologic Evidence for Neck, Upper Extremity and Low Back.* Second printing. U.S. Department of Health and Human Services.

Nogareda Cuixart, S. et al. 2011. Evaluación del riesgo por manipulación manual de pacientes: Método MAPO. Instituto Nacional de Seguridad e Higiene en el Trabajo. NTP no. 907.

NRC. 2001. *Musculoskeletal Disorders and the Workplace: Low Back and Upper Extremity.* Washington, DC: National Academy Press.

Occhipinti, E., Colombini, D., Molteni, G., Menoni, O., Boccardi, S., Grieco, A. 1989. Clinical and functional examination of the spine in working communities: Occurrence of alterations in the male control group. *Clinical Biomechanics* 4(1): 25–33.

OSHA 2000 fact sheet 10. 2000. Work Related Low Back Disorders. https://osha.europa.eu/en/publications/factsheets/10.

OSHA 2007. *A Back Injury Prevention Guide for Health Care Providers.* http://www.dir.ca.gov/dosh/dosh_publications/backinj.pdf

Ottenga, F., Cristaudo, A., Guidi, M., Paladino, G., Giuliano, G., Serretti, N., Talini, D. 2002. Phenomenon of occupational accidents of workers of IRCCS of neurosciences. *Giornale Italiano de Medicina del Lavoro ed Ergonomia* 24(2): 131–137. (Italian)

Owen, B. D. 1989. The magnitude of low-back problem in nursing. *Western Journal of Nursing Research* 11(2): 234–42.

———. 2000. Preventing injuries using an ergonomic approach. *Association of periOperative Registered Nurses Journal* 72(6): 1031–1036.

Owen, B., Garg, A., Jensen, C. 1992. Four methods for identification of most back-stressing tasks performed by nursing assistants in nursing homes. *International Journal of Industrial Ergonomics* 9, 213–220.

Palmer, K. T., Walker-Bone, K., Griffin, M. J., Syddall, H., Pannett, B., Coggon, D., Cooper, C. 2001. Prevalence and occupational associations of neck pain in the British population. *Scandinavian Journal of Work and Environmental Health* 27(1): 49–56.

Plouvier, S., Leclerc, A., Chastang, J. F., Bonenfant, S., Golberg, M. 2009. Socioeconomic position and low-back pain—The role of biomechanical strains and psychosocial work factors in the Gazel cohort. *Scandinavian Journal of Work and Environmental Health* 35(6): 429–436.

PNLG9 (Italy). 2008. http://www.isico.it/isico/pdf/LineeGuidaErniaDisco.pdf

Punnett, L., Prüss-Utün, A., Nelson, D. I., Fingerhut, M. A., Leigh, J., Tak, S., Phillips, S. 2005. Estimating the global burden of low back pain attributable to combined occupational exposures. *American Journal of Industrial Medicine* 48(6): 459–469.

RCN (Royal College of Nursing). 2001. *Manual handling assessments in hospitals and in the community. An RCN guide.* London: The Royal College of Nursing.
———. 2003a. *Safer staff, better care.* London, The Royal College of Nursing.
Seidler, A. et al. 2009. Cumulative occupational lumbar load and lumbar disc disease—Results of a German multi-center case-control study (EPILIFT). *BMC Musculoskeletal Disorders* 10: 48.
Silverstein, B. A., Bao, S. S., Fan, Z. J., Howard, N., Smith, C., Spielholz, P., Bonauto, D., Viikari-Juntura, E. 2008. Rotator cuff syndrome: Personal, work-related psychosocial and physical load factors. *Journal of Occupational and Environmental Medicine* 50(9): 1062–1076.
Smedley, J, Inskip, H., Cooper, C., Coggon, D. 1998. Natural history of low back pain. A longitudinal study in nurses. *Spine* 23(22): 2422–2426.
Smedley, J., Inskip, H., Trevelyan, F., Buckle, P., Cooper, C., Coggon, D. 2003. Risk factors for incident neck and shoulder pain in hospital nurses. *Occupational and Environmental Medicine* 60(11): 864–869.
Smith, D. R., Sato, M., Miyajima, T., Mizutani, T., Yamagata, Z. 2003. Musculoskeletal disorders self-reported by female nursing students in central Japan: A complete cross-sectional survey. *International Journal of Nursing Studies* 40(7): 725–729.
Squadroni, R., Barbini, N. 2003. Ergonomic analysis of nursing activities in relation to the development of musculoskeletal disorders. *Assist Inferm Ric* 22(3): 151–158.
St. Vincent, M., Tellier, C., Lortie, M. 1989. Training in handling: An evaluative study. *Ergonomics* 32(2): 191–210.
Stobbe, T. J. et al. 1988. Incidence of low back injuries among nursing personnel as a function of patient lifting frequency. *Journal of Safety Research* 19: 21–28.
Stubbs, D. A., Buckle, P. W., Hudson, M. P., Rivers, P. M. 1983. Back pain in the nursing profession. II. The effectiveness of training. *Ergonomics* 26(8): 767–779.
Stubbs, D. A., Buckle, P. W., Hudson, M. P., Rivers, P. M., Baty, D. 1986. Backing out: Nurse wastage associated with back pain. *International Journal of Nursing Studies* 23(4): 325–336.
Toivanen, H., Helin, P., Hanninen, O. 1993. Impact of regular relaxation training and psychosocial working factors on neck-shoulder tension and absenteeism in hospital cleaners. *Journal of Occupational Medicine* 35: 1123–1130.
Trinkoff, A. M., Le, R., Geiger-Brown, J., Lipscomb, J., Lang, G. 2006. Longitudinal relationship of work hours, mandatory overtime, and on-call to musculoskeletal problems in nurses. *American Journal of Industrial Medicine* 49(11): 964–971.
Ulin, S. S., Chaffin, D. B., Patellos, C. L., Blitz, S. G., Emerick, C. A., Lundy, F., Misher, L. 1997. A biomechanical analysis of methods used for transferring totally dependent patients. *SCI Nursing* 14(1): 19–27.
U.S. Department of Labor, Bureau of Labor Statistics. 2000. Table R6: Incidence Rates for Nonfatal Occupational Injuries and Illness Involving Days Away from Work per 10,000 Full-Time Workers by Industry and Selected Parts of Body Affected by Injuries or Illness.
Van der Windt, D. A. W. M., Thomas, E., Pope, D. P., de Winter, A. F., Macfarlane, G. Y., Bouter, L. M., Silman, A. J. 2000. Occupational risk factors for shoulder pain: A systematic review. *Occupational and Environmental Medicine* 57: 433–442.
Vasseljen, O., Holte, K. A., Westgaard, R. H. 2001. Shoulder and neck complaints in customer relations: Individual risk factors and perceived exposures at work. *Ergonomics* 44(4): 355–372.
Videman, T., Rauhala, H., Asp, S., Lindström, K., Cedercreutz, G., Kämppi, M., Tola, S., Troup, J. D. 1989. Patient-handling skill, back injuries, and back pain. An intervention study in nursing. *Spine* 14(2): 148–156.

Viikari-Juntura, E. 2001. Limited evidence for conservative treatment methods for work-related neck and upper-limb disorders—Should we be worried? *Scandinavian Journal of Work and Environmental Health* 27(5): 297–298.

Violante, F. S., Fiori, M., Fiorentini, C., Risi, A., Garagnani, G., Bonfiglioli, R., Mattioli, S. 2004. Associations of psychosocial and individual factors with three different categories of back disorder among nursing staff. *Journal of Occupational Health* 46(2): 100–108.

Waters, T., Collins, J., Galinsky, T., Caruso, C. 2006. NIOSH research efforts to prevent musculoskeletal disorders in the healthcare industry. *Orthopaedic Nursing* 25(6): 380–389.

Waters, T. R., Nelson, A., Proctor, C. 2006. Patient handling tasks with high risk for musculoskeletal disorders in critical care. *Critical Care Nursing Clinics of North America* 19: 131–143.

Winkelmolen, G. H., Landeweerd, J. A., Drost, M. R. 1994. An evaluation of patient lifting techniques. *Ergonomics* 37(5): 921–932.

World Health Organization, WHO. 2011. World Health Statistics, 115–124. Retrieved form www.who.int/whosis/whostat/2011/en/

Yassi, A., Khokhar, J., Tate, R., Cooper, J., Snow, C., Vallentyne, S. 1995. The epidemiology of back injuries in nurses at a large Canadian tertiary care hospital: Implication for prevention. *Occupational Medicine* 45(4): 215–220.

Yassi, A. et al. 2001. *Basic Environmental Health.* New York: Oxford University Press.

Yip, Y. B., Ho, S. C., Chan, S. G. 2001. Socio-psychological stressor as a risk factors for low back pain in Chinese middle-aged women. *Journal of Advanced Nursing* 36(3): 409–416.

# Index